The Water Keeper's Handbook

THE
WATER
KEEPER'S
HANDBOOK

Michael J. Robbins

The Crowood Press

First published in Great Britain by
The Crowood Press
Ramsbury, Marlborough,
Wiltshire SN8 2HE

British Library Cataloguing in Publication Data

Robbins, M.J.
  The water keeper's handbook
  1. Angling waters: trout fisheries. Management
  - Manuals
  I. Title
  799.1'755

  ISBN 1 85223 117 3

To David Clarke, with thanks for his help and encouragement.

Picture Credits:

Photographs by Jim Tyree
Line illustrations by Audrey Robbins
Cover design by Pat Warren

Typeset by Columns of Reading
Printed in Great Britain at the University Printing House, Oxford

Contents

Acknowledgements

Fishing is undoubtedly an illness, congenital in some cases, but contracted later in life in others. Its symptoms vary but, in the initial stages, the sufferer is usually in a state of high fever. Most victims experience some sort of remission and the unlucky ones may even be cured, but a few remain thus afflicted for the rest of their lives. Having long since left the feverish state I still love fishing, but now I am just as happy trying to improve the environment for both fish and fishermen. The hunting instinct is still there, but it is mixed with the urge to conserve and to put something back into a sport which gives so much pleasure. Most of us, both hunters and conservationists, are actually paddling the same canoe, and the sooner we realise that we have a common interest at heart and work together against the polluters the more successful we shall be in our aims. Only the pressure of united public opinion will bring results.

I claim no originality for the ideas expressed in this book; most of them can be found elsewhere, though sometimes the work has only been done on an experimental basis. What I can say, however, is that we have actually *done* all these things, with varying degrees of success. My aim has been to describe how the different jobs were tackled, so that you may avoid some of the disasters which came our way.

I make no apologies for writing this book and I will attach the blame to those who deserve it. The main culprit must be David Clarke, whose idea it was that our branch of the Salmon and Trout Association should provide trout fishing, in order to increase the membership. Not only did he provide the idea, he also provided the river and a large amount of material help as well. A smaller portion of blame goes to my wife, who would not let me refuse to write this book, though as she has done the drawings I suppose that cancels out her share of blame. Jack Fitt was too modest to contribute the section on stillwater fisheries, but much of the credit for this part of the book must go to him.

Acknowledgements

Unqualified thanks go to Pat Gibbons. Having served a long apprenticeship on my handwriting by typing all my fishing club letters, she has typed the manuscript and tried to put some discipline into my punctuation. Pat and her husband Roger have been a great help to me in all my fishing club work and I am very grateful to them. Finally, thanks to Jim Tyree who has taken all the photographs; his patience has been unending and his ideas a great help. This book has been a team effort and my gratitude goes not only to those I have just mentioned, or blamed, but also to all those club members who have helped with the work on the different fisheries. May you, in your turn, have such loyal and enthusiastic support.

1
All Change!

When I first moved to Norfolk about thirty years ago, fishing was very difficult, because you actually had to push the fish out of the way before you could get the bait into the water; in fact, fish were so abundant that water levels were about a foot higher than they are today. Those were the days before the 'agricultural revolution', the days before pesticides, herbicides and nitrates had begun to take their toll. Gradually the numbers of young fish lessened and, year by year, the population diminished until by the late 1970s few fish remained in what was still classified as pure water. It's no use crying over spilt milk, however. Just as industrial pollution has been substantially checked, I am sure that agricultural practices will gradually improve, but it will be many years before the rivers of Norfolk recover fully.

As far as my local rivers are concerned the decline in the coarse fish population has made the creation of trout fisheries very much easier, especially as now there seems to be nothing in the water which is harmful to adult fish. Although carp and pike seem to be the fashionable fish at present, there is still a great demand for trout fishing and, in the case of rivers, this is increasing. In fact, the last twenty-five years have seen enormous changes in almost every aspect of our sport and in the case of tackle, there has been a revolution. I'm only in my fifties, yet I remember in my teens when plaited silk lines and gut casts were standard equipment for those who fly fished. However, trout fisheries are my brief and they provide yet another example of the changes which have taken place. Most people who lived in the lowlands of Britain used to consider trout fishing the province of a fairly affluent minority. However, when Chew Valley Lake opened, trout fishing became more readily accessible, and it produced the first trickle of converts. Grafham Water opened the floodgates and vast numbers of coarse fishermen turned to trout fishing, if only during the close season for coarse fish. Many smaller lake fisheries

Fig 1   *The upper reaches –*
easy to work in.

Fig 2   *Natural trout water,*
but nature appreciates a
helping hand.

were opened to take advantage of this boom, though some of them were very expensive and exclusive. It seems that the demand for smaller still waters has levelled out, but many of the new breed of trout fishermen are keen to enjoy the pleasures of river fishing as well.

There are very many small lowland rivers which could make useful trout fisheries or where the existing trout fishing could be vastly improved. I used to live in Bristol and, even in those days, several tributaries of the Avon provided some trout fishing, and many more could have done so. The same opportunities are present in many parts of Britain. Some of the Thames tributaries, and those of the Trent, already provide good trout fishing. A bit of industry and self-help on a river can work wonders.

One more change needs to be considered at the moment. Privatisation of Regional Water Authorities seems imminent, in which case a National Rivers Authority (NRA) will come into being. If this happens, I hope that the NRA will have the teeth, the financial power and, above all, the *will* to prosecute the polluters. Anyway, though the titles may change, there will still be officers to approach, when you need help and advice. For many of us, to have a National Rivers Authority will be no real change, as we started fishing in the old days of the River Boards and we may possibly be glad to see something like it come back again. The present system, where the water authority is supposed to enforce the legislation on pollution, but can often be the polluter, just does not work.

In this book I am going to deal with small, lowland rivers, not the more natural trout waters, most of which are found in hillier regions. As for lakes, they are most likely to be old gravel pits; these are the waters which are most often available, and new ones are being excavated all the time. The first step when wishing to create or improve a trout fishery is to acquire or rent some water in which trout will actually live. This is probably the biggest hurdle of all, as competition for water is keen, especially in the more populous areas. Established clubs are in a happier position, since the larger ones often rent or own many waters, and it is merely a question of development or change of use (not that I would ever advocate trying to change a good coarse fishery into a trout water).

If you are not lucky enough to control some water already, don't despair. Our local branch of the Salmon and Trout Association rents five waters, in different parts of Norfolk, yet less than ten years ago we had no fishing at all. It is important to find the right person to negotiate for you; our advantage is that some of the members are farmers and landowners and so, directly or indirectly, they know people who have lakes and rivers on their land, even if they haven't any fishing rights themselves. Farmers are often wary of strangers, and with good reason; there seems to be an ever increasing number of idiots who leave gates open, drop litter and let their dogs roam freely. Farmers will be much more ready to negotiate with somebody they know, and there is no doubt that our association's farmer members have made the renting of water much easier for us. Where gravel pits are concerned, it is possible that friends in the building trade may be useful, especially if they have business dealings with the firms that are extracting the gravel. All this may sound a little devious, but you must give yourself the best possible chance of success, and in our case it really has paid handsome dividends.

Once you have found a water that you may be able to rent, ask the Fisheries Officer from the Water Authority to test it for you. There's no point in renting a water, however pretty it may look, if the trout come floating, belly up, as soon as they are put in. You certainly need to know that the water quality, the pH and the food supply are all good enough

Fig 3   About the optimum size of river to work on.

to give you a reasonable chance of success. You should also be aware that there is a world of difference between a water seen in mid-summer and in mid-winter. A nice, open lake in mid-winter may well be choked right to the surface with weed six months later. It is easy enough to test the depth of a lake merely by using coarse fishing gear, float and plummet. If there are places near the bank with four or five metres (13–16 feet) of water there should always be some areas of open water, even in the weediest of years. Even so, it is highly desirable that a lake should have shallow areas as well, as variety in the habitat is a great asset. Similarly, it pays to do a little surveying work on a river. At its simplest, take a long landing net handle and poke about to see what the river bed is like. It will almost certainly be covered in mud but, if you are lucky enough to find gravel underneath, that is a bonus. Gravel is not essential but it does improve your prospects. It's also useful to have a picture of the depth of the river; too deep and it may not provide very good fly fishing; too shallow and there may be a lack of fish holding places, though this last snag is not too difficult to remedy. Finally, use your imagination a bit, only don't let it run riot! Would you like to fish

Fig 4   The lower reaches – ideal for improvement.

the water? If the answer is 'yes' and the water quality is suitable, then go ahead with your negotiations.

It is most unlikely that the owner will agree to any long-term arrangement in the first instance. He will want to see whether your 'mob' are suitable tenants! This is unfortunate, as you are probably going to do quite a lot of hard work to begin with, and it would be very disappointing to see the fruits of your labours pass rapidly to someone else. This possibility involves a question of mutual trust, and is a chance that you will have to take. It is wise to have professional advice when drawing up an agreement. You need a *licence* to fish; the word 'lease' should be avoided like the plague, as that will almost certainly land you with the need to pay sporting rates. I believe that Savill's, for example have a form of agreement which is suitable but, wherever you go for advice, this is certainly not a 'do-it-yourself' job, and you will probably have to pay all the legal costs involved for both sides! If there are several owners involved, as may well be the case on a length of river, the complications multiply. We had this problem on one fishery, but the owner of the longest stretch helped in the negotiations with the others.

Your professional adviser should give you an idea of what constitutes a fair rent, as this is an important part of the agreement. A clause to review this rent regularly and to increase it, at least in line with inflation, may help with your security of tenure. The owner will almost certainly want to retain the right of one or two rods, so that he or his friends may be able to fish. These anglers would normally abide by the rules and limits of the club. Other important issues are access and parking – a farmer has work to do and doesn't want to find cars blocking his way. Finally, there is the question of what you will be allowed to do to improve the water. It is vital to have a precise understanding between club and owner on bank clearance and levelling, the felling and planting of trees and the use of instream devices to increase the fish holding potential of the water. The importance of conservation is gradually beginning to be appreciated and if the landowner becomes interested and involved, many would-be problems will simply evaporate. The most important thing is to build up a good relationship with the owner, for this is a partnership. I know that a couple of *our* owners are highly delighted that their waters have developed into such good trout fisheries. So, sign your agreement and I wish you luck!

2
River Improvements

If the initial work has been successful, and negotiations have reached a happy conclusion, you should now have control of a mile or more of both banks of a river. You may even have the more or less enthusiastic support of the landowners, some of whom are perhaps misguided enough to fish themselves. Apart from the goodwill of the owners, you need a few friends who don't mind getting their hands dirty; you cannot do all the work by yourself, nor should you. Helping with the work, rather like tying your own flies or making your own tackle, adds an extra dimension to the pleasure you get from your fishing. In any club some members will be able to contribute very little in the way of work; they may be old or unwell or have too many other commitments. Others are just not interested; they have plenty of time to fish but never to work. I feel rather sorry for them, as they miss a lot of enjoyment. Working parties are fun and help to foster a good club spirit. I will return to the subject of working parties when discussing the running of a club (chapter 8), but now you are ready to start work – almost.

Before you can really get wet, you may need a boat. In fact, if you cannot wade almost the whole of your stretch of river in chest waders, without imminent danger of shipping water, you will certainly need a boat. The ideal craft is quite a big punt, because a stable platform is required from which to work. To try to swing a sledge hammer whilst standing up in a fibreglass dinghy invites speedy immersion of both hammer and operator, and hammers are expensive! I am lucky enough to have a large aluminium punt, generously provided by the same benefactor as provided my club's river. This is a perfect craft to work from; one can step right on to the side and it barely tilts. Two or three people can easily work in a boat like this at the same time, without fear of accidents, and Fig 5 gives an accurate impression of both its size and stability. Its actual dimensions are 12ft long by 5ft wide; it declines to be metric! The only warning I would give is that you must have a non-slip

Fig 5   The ideal punt. Robert is helping me to put the fish stop in place.

floor to stand on. I have stapled wire netting to the floor-boards of our punt and this is completely satisfactory. Unfortunately, a punt like this costs a lot of money, especially if you have to buy a new one. Even so, you will have to beg, borrow or steal a suitable craft for the job. You might even ask if your water authority can help. As well as serving as the platform from which to work, the boat will also carry all your tools and materials to where you need them, a great saving in energy and time. It is surprising how often a boat is needed, especially in the early years when there is a lot of improvement work to be done, so it is essential to organise something. Those whose river is shallower will certainly find work much simpler. A couple of pairs of chest waders are much cheaper than a punt, but make sure that you buy strong ones and, even then, beware of barbed wire!

At last you are in a position to start work – but where? Be quite clear what your aims are and have a set of priorities. Haphazard messing about will not achieve a great deal. Basically, the biggest problems in lowland rivers are likely to be weeds and mud. Because the river has probably been neglected, fish stocks are likely to be quite low, and trout almost non-existent. There is no point in tidying up the banks for the benefit of the fisherman if there are no fish for him to catch. You could simply tip in some trout – I wouldn't call this stocking – and a few of them may be caught very soon after release, in the most favoured bits of water, but most of them will simply disappear. It's obvious, therefore, that you must look after the fish before looking after the fisherman.

I don't suppose that a good home for the fish differs all that much from our basic requirements, food and shelter, or in the case of the fish probably the other way around, as they seem to accept quite short rations provided that they feel secure. The old saying 'variety is the spice of life' is very applicable to managing the river; strive to avoid uniformity of habitat. You need deeps and shallows, bends and straights, shaded water and open stretches, silt and gravel. Forget the trout for a moment and consider their food. Some insects like gravel, others live on the undersides of stones, others, mayfly nymphs for example, prefer to live in silt, and yet others will live on and among the weeds. Many fine stretches of river have been wrecked by badly planned dredging, which has left in its wake a featureless canal – and one which silts up even quicker than before it was dredged! You must strive to create a mixed environment, rich in its variety of forms of life. If you succeed in doing this, the fish will find somewhere to hide, somewhere to feed, and a reasonably well-stocked larder.

Fishery research, both in Britain and in America, has shown that a little work in and around the river can bring dramatic improvements in the levels of fish stock which a river can carry. The devices most commonly used to effect this improvement are groynes (current deflectors), and low weirs. In other experiments, rafts have been moored to the bankside to provide overhead cover, though natural cover of this sort comes from undercut banks and from overhanging trees and bushes. It is certain, however, that overhead cover of any sort is most attractive to the fish. In one case, this alone has more than doubled the number of fish present in a stretch of river, and this was a stretch which was already quite well populated. On the stretch of river which I have looked after longest, the catch of trout rose from just over 100 trout a season during the first two years to over 500 by the fifth year, and it seems to be holding steady at that level. The number of fish stocked has been slightly increased, but is in no way proportional to the increase in catch. I am certain that this improvement is a direct result of the efforts made to upgrade the condition of the river bed, and the planting of bankside

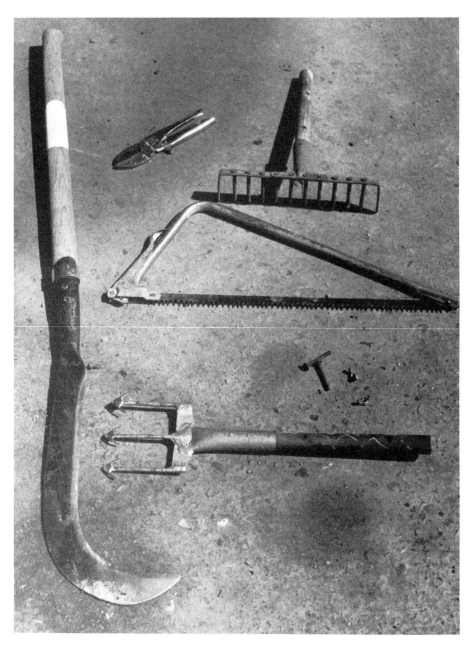

Fig 6 *A few tools of the trade. The trident is for spearing pike and is mounted on a very long aluminium pole.*

cover. Results like this give us plenty of encouragement to carry on with the good work. I hope that you will be encouraged as well, but don't expect instant results – it wasn't until our third season that we began to enjoy improved fishing.

MATERIALS NEEDED

Many of the jobs which will be tackled need fairly similar materials, so a check-list may be useful:

Wire netting	25mm or 31mm mesh.
Angle irons	
Poles	10cm to 15cm diameter. Larch are the very best, but any straight ones will do.
Staples	Not the smallest size, about 20mm to 25mm will do nicely.
Wire	A good strong tying wire. People who put up television aerials have plenty of scrap.

For this sort of work, the term 'angle irons' covers a multitude of sins and you grab anything that may be available at the right price. Old scaffold poles are excellent, the angles off an old bedstead do very well. Wooden posts may be used, but they are not as durable. A good supply of these materials for uprights is needed, between one and three metres (3–10 feet) long, depending on the depth of the river.

It's useful to have a bag of tools handy, containing wire-cutters, a little hacksaw, hammer, thick gloves, secateurs, nails, baler-twine: for bigger tools, sledge-hammer, spade and saw are essentials.

GROYNES

The simplest device for improving a river's fish-holding capacity, and probably the most effective one, is the current deflector or groyne. The ones I use consist of a horizontal pole to which wire netting has been firmly stapled. One end of the pole is secured to the river bank (I find it best to dig that end in) and angle irons are used to hold it in place out in the river, at the required angle to the current. This pattern of groyne is almost certainly the cheapest and easiest to make. If you find that you

Fig 7   *Materials needed to make a groyne.*

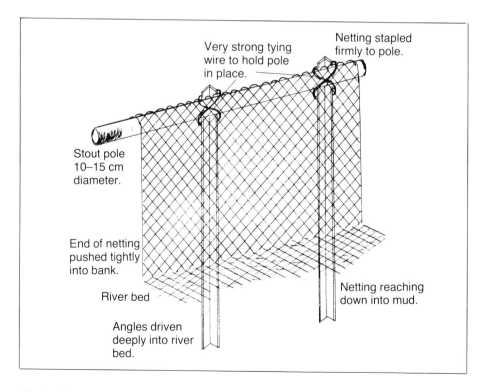

Very strong tying wire to hold pole in place.

Netting stapled firmly to pole.

Stout pole 10–15 cm diameter.

End of netting pushed tightly into bank.

River bed

Angles driven deeply into river bed.

Netting reaching down into mud.

Fig 8   *Basic construction of a groyne.*

have sited it in the wrong place, it is not very difficult to take out. The angle irons can present a problem, as quite a strong upward pull is needed to lift them out – I usually devise some sort of lever to do this. Don't use wire netting with too wide a mesh – 31mm mesh is about right, certainly not larger. Within a couple of days the netting will be completely clogged by the debris brought down by the current, thus becoming a solid barrier and deflecting the current. Until you have tried this, it is hard to believe just how much debris is floating down what appears to be a crystal clear river. You can, of course, use stout poles instead of angle irons for the uprights. They will be a little cheaper, but they are much harder to hammer through the river bed and they will not be quite as strong and durable.

The groyne should stand only a few centimetres above normal water level, so that flood water will pass straight over the top. The end at the bankside should also be several centimetres higher than the end which is out in the river. The side of the wire netting is pushed into the bank as firmly as possible. The bottom of the wire netting – which should be appreciably longer than the river is deep – is pushed right down into the mud with an oar and more mud pulled on to the top with a chrome or similar implement (*see* Fig 9).

Fig 9   A chrome; most useful for pulling out mud.

21

I must emphasise the need to make the groyne really strong, especially if the water is a metre (3 feet) or more deep. The weight of water in times of flood is considerable, and to make a flimsy construction is a waste of time, effort and materials. A groyne three metres (10 feet) long will need two angle irons to support it. Obviously, the longer the groyne, the more angle irons are needed, not only to support the weight but also to prevent the wire netting bowing out in the current.

I have also made groynes by driving a line of wooden stakes into the bed of the river, as shown in Fig 10. These may look a little better than the other sort, once the tops have been trimmed off with the chain saw, but they are much more expensive. It is also extremely hard work to hammer home the large number of stakes needed, especially if they have to go down into a bed of gravel which you are hoping to expose. Long before we started to look after our river, the old River Board made three experimental groynes, using heavy asbestos sheets. These seem to do the job well, but they are no better to look at, and much more labour is needed to get such a heavy construction in place. In rocky rivers, some groynes are made by putting stones and flints into strong wire baskets, thus building a wall of stone. I have only seen these on salmon rivers, and the materials used in their construction would not be readily available on a lowland river. I strongly advocate the use of the pole and wire netting groynes that I have illustrated. They seem to have all the advantages for our type of water. They are cheaper, lighter, easier to make, and the netting adapts to the varying contours of the river bed.

Fig 10   Not my type of groyne – too expensive and too much like hard work.

Positioning of Groynes

Before any groynes are put in place, you should consider exactly what their function is going to be. Are they to provide small pockets of shelter, which may prove attractive to a fish or two or, at the other extreme, are they to deal with a badly silted stretch of river which needs drastic treatment? Many rivers have been dredged and widened to such an extent that the speed of the current is no longer fast enough to carry silt away. Abstraction, too, may have reduced the volume of water flowing down a river.

The following illustrations suggest some of the uses to which groynes may be put. Common sense, and the making of a few mistakes, will soon help you to decide what sort of treatment is best suited to your river. You will notice that the groynes are set at about 45 degrees to the bank, which makes them more able to withstand the weight of the current. I certainly made some mistakes early on by placing groynes at right angles to the bank. I then discovered that angle irons would bend surprisingly easily! As a general guide, the faster the river flows, the smaller the groynes are likely to be.

Fig 11   A double groyne at the down stream end of a shallow.

A straight in which the gravel bed has been covered by silt (Figs 12 and 13). There may be up to a metre (3 feet) of black, evil-smelling mud in such a river, which is no use to anybody. The gap through which the water flows should be suited to the depth of silt and the speed of the current. On a really poor, sluggish straight, the gap between the ends of the groynes might be no more than one-third of the total width of the river. Remember, the groynes should only stand a few centimetres above normal water-level, so that flood water will pass straight over the top. Your aim is to clear the silt, not to flood the whole county! If you find that you have left too narrow a gap, saw the last 50cm (20in) off the end

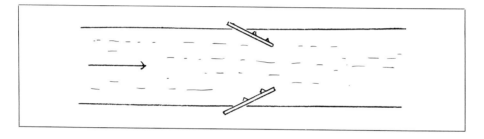

Fig 12   Double groyne on a straight stretch of river.

Fig 13   A pair of double groynes at work. The acceleration of the current is clearly visible.

of each groyne and fold the wire netting back – don't, therefore, put the outermost angle iron right at the end of the groyne. If the gap is too wide, it is possible to add another small pole, but this never looks very tidy, it just looks like the afterthought it is.

A groyne of this pattern, once it has been in place for about a year, should have cleared out a lovely midstream run twenty or thirty metres (20–30 yards) long, exposing the natural bed of the river. This ought to give the fish quite a deep hole in which to hide, a nice streamy bit of water carrying their food to them and, in times of flood, a bit of shelter in the slack water behind the groynes.

A straight in which the main aim is to improve the river's fish holding capacity (Fig 14). Single groynes are built on alternate sides of the river, thus sending the current to and fro and producing more variety. I have used groynes like this, and trout lie near them, but I have not found this pattern especially useful. Mud is my biggest enemy and I find that the double groyne is more effective on the stretch of river that I look after.

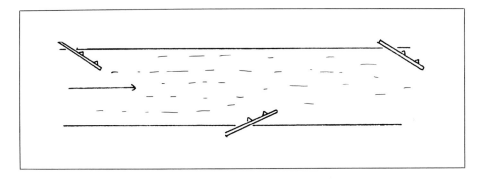

Fig 14   Groynes to send current to and fro.

A straight in which the main flow is down one side of the river and the other bank is silting up (Fig 15). I have found it useful to place all the groynes on the side which is silting up. This helps to keep the main channel silt-free and seems to allow the silted side to build up more quickly and become firm bank. The far bank will provide some good feeding stations for the fish, especially if it is well bushed. It is important not to make this sort of groyne too big or the end result will be rapid erosion of the far bank.

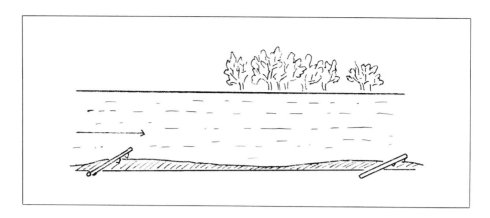

Fig 15 *Groynes to push current to one side of the river.*

To speed the current round the outside of a bend (Figs 16 and 17). This type of groyne, placed on the inside of the bend, has proved extremely successful for me, especially when the outside of the bend has good bushes growing to give extra cover to the fish. As with the previous example, it is important not to make this groyne too big, or you will merely cause rapid erosion of the outside of the bend, thus nullifying any improvement that may have been made initially. I tend to make a groyne

Fig 16 *Groyne to speed current round the outside of a bend.*

*Fig 17   There is a groyne on this bend, hidden by the
rushes. It pushes the current under the willows on the far
bank – a super fish holding spot.*

like this rather small in the first instance and, contrary to my earlier advice, I add a bit to the end if it is not making sufficient impact.

A bend such as this is illustrated in Fig 17. An added feature here was that the opposite bank from the groyne, just where the bend was beginning, had been very badly eroded, leaving quite a large area of swamp. This has all been reclaimed, in the manner described later in this chapter, and is now good firm ground again.

Mini groynes (Figs 18 and 19). I have been experimenting with small groynes which only extend about a metre (3 feet) out in the river. I have stretches of river which are very short of trees and cover, and where farm animals are a nuisance, and these mini groynes are intended to provide some cover and add some variety. All the poles are of green willow and are pushed well into the bank, at least 25cm above normal water level. Only the middle pole has a bit of wire netting stapled to it, more to cause a bit of turbulence than anything else. The poles will all take root and grow, and some of the vegetation will be out of the reach of animals, unless the farmer changes from sheep to giraffes! These particular groynes have been in use for a couple of years and seem to be beneficial, but it is too early to make a positive claim of efficacy.

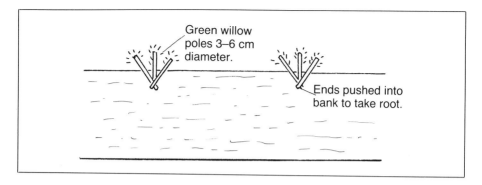

Fig 18   Mini groynes.

*Fig 19   A mini groyne. The willows are beginning to grow,
out of reach of the animals.*

Disadvantages

Do not be deceived by my enthusiasm for the use of groynes. If I could
do without them, I most certainly would! Though they do a lot of good
work, they are not a panacea. The first obvious objection is an aesthetic
one. Groynes, however neatly they have been made, are certainly not
things of beauty, though their effectiveness will be a source of joy for

ever, or very nearly so, if they have been strongly built. It is possible to distract the eye by planting a bush or two on the bank by them, or by hammering a long willow branch into the bank, leaving a couple of metres protruding and securing the branch to the pole of the groyne. The willow stake is certain to take root and will soon provide an extra bit of bankside cover which, on the upstream side of the groyne, sometimes proves attractive to a trout. The eddy is very often the home of a resting pike and we have shot or speared several in such a place. Sometimes I have made the pole of the groyne from a thick willow branch, if I can find one straight enough. One such groyne is shown in Fig 21, but big, thick willow branches such as this one do not strike as easily as thinner ones.

The most serious disadvantage of groynes, however, is that though they clear silt away they also build it up — great big banks of the stuff! Inevitably, the back eddy encourages the depositing of silt, and a large bank soon builds up behind each groyne, some worse than others depending on the set of the current. I can never predict exactly what will happen, except that there will be a clear, deep run in the main current,

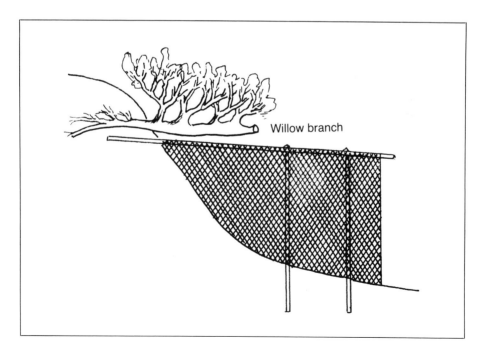

Willow branch

Fig 20 Use of willow cutting to help disguise a groyne.

and I think the advantage of this far outweighs the two silted-up eddies. Where the groyne has been sited on a bend to speed the current round the outside, it only serves to hasten a process which is already happening. Silt settles on the inside of the bend, as the current there is slower, and the bend makes a natural eddy.

Despite these disadvantages, groynes do a great deal of good, especially in rivers where, for one reason or another, the speed of the current is no longer sufficient to keep the river bed reasonably clear of silt. It is important not to make the groynes so big that they cause problems with erosion, nor should there be so many of them that they start to look ridiculous. The building and use of groynes calls for little more than a bit of common sense. Electro-fishing has shown a much greater abundance of fish near the groynes than in other stretches of the river I look after. This applies not only to the trout; we have good shoals of dace, and the best ones are always near the groynes. I am sure that some coarse fisheries would benefit in the same way as we have. Roach, dace, chub and grayling should all favour the streamy runs provided by pairs of groynes and this would be ideal water for long-trotting.

Fig 21   A groyne made from a willow branch, which is growing well.

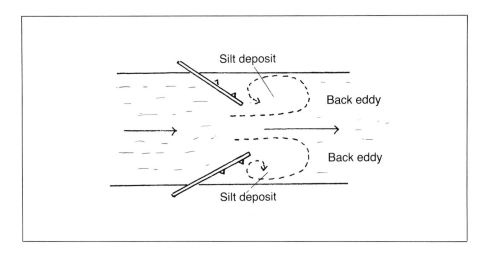

Fig 22   Build-up of silt in back eddies.

LOW WEIRS

Weirs form a very attractive and fishy feature of any river, a constant source of lovely, well-oxygenated water. The weir pool compels the angler to make at least fifty 'last casts', hoping to catch its legendary 'monster'. As you can see, I am very much in favour of weirs. Before building one, be absolutely certain that it has a purpose and that it is not just a folly, pleasing to the eye and ear but doing little to improve the fish holding capacity of the water.

If your water is much larger than a small river, I doubt whether you have any genuine need to build a weir, and I also doubt whether your river authority would give you permission to build one in any case. Apart from this, the bigger the water, the bigger the undertaking and, even on a small water, building a decent weir is hard work and not something to be embarked upon lightly. Water has a very strong objection to having its level raised and will find every possible excuse not to co-operate; it will find a way round the sides, or even scour a channel underneath if it can. No weir which you build should aim to raise the water level very much – 25 to 30cm (10–12in) at the most. In a lowland river, I doubt whether there will be a fall of much more than a metre (three feet) over a length of two or even three kilometres (one or two miles) of river. If you are lucky enough to have a steeper gradient than that, you should certainly have a nice brisk current and plenty of opportunities to let the water do some work for you. Actually, weirs are often of more use on the faster-flowing streams, to provide holding areas of deeper water. Some of the rivers in Derbyshire, for example, often contain a succession of stone dams, which are both effective and pleasant to look at.

This leads me to the only really compelling reason to install a weir. In your river there may be a very long shallow, which does not hold fish as well as it might. In a case like this, an extra 20cm (8in) of water can make a very big difference to the fish holding potential, without slowing down the current so much that all you are making is a settling pond for the silt. If there is such a site near the upstream end of your water where a weir may be beneficial, the weir will then have another use – it will discourage (I didn't say stop) your stock fish from running up into your neighbour's water. Trout are expensive, so there is no point in providing others with stock fish free of charge, if you can avoid it. Regrettably, the converse does not apply. Your stock fish are quite likely to take an unholy delight in charging off downstream over a weir. On one fishery we stocked several hundred metres upstream of a mill pool and, within a few hours, the pool had numerous new tenants! Fortunately for us, the mill pool is part of our fishing and only about half-way down our stretch of water, so these trout were not lost to us, at least, not immediately; the pool is next to a little road, and is a favourite poaching place for the local lads.

I have built two weirs so far, and am in the throes of making the third (*see* page 39). The first one was a mistake; I swiftly took the middle out leaving a very effective double groyne. Fortunately, this cleared a splendid stretch of gravel which the trout like very much, so the labour was not entirely wasted. In self-defence, I was very 'green' indeed when that weir was built but, as I keep stressing, if you are afraid of making a few mistakes, you won't *make* anything. The most important thing is to learn from the mistakes.

The second weir, shown in Fig 23, is further upstream on the same river, where it is quite narrow. This weir is older, and has proved much more successful. I wished to tackle a straight stretch of river, this one being almost 200m (220yds) long. The upstream end was beautiful gravel, but very shallow. The bottom part tended to be mostly sandy, featureless and, apart from a few small trout, fishless. We decided to try a weir about half-way down this stretch, raising the water at that point by about 25cm (10in). The results have been very interesting; the stretch upstream of the weir is very much more to the trouts' liking now that the water is a bit deeper. A good number of fish now lie along the whole

*Fig 23   A pretty weir, which also narrows the river
considerably.*

stretch to the weir, in fact the sequence of photographs on page 168, showing that the author does actually catch a fish now and again, is taken at the upstream end of this shallow. The stretch downstream of the weir only serves to illustrate how stupid trout are, and that fish don't know what is good for them. There is quite a nice little weir pool and, for another 30 to 40 metres (33–44 yards), much of the sand has been cleared away, leaving some lovely gravelly runs. Despite what would seem to be obvious attractions, the trout population on this stretch has not yet increased significantly. One of our trout suppliers, who also runs a river fishery, commented to me one day that he had long given up forecasting what might or might not be good lies for trout; the only good spots are those which the trout actually choose themselves to frequent, and what may look a 'desirable residence' for a trout to us will often fail to find a buyer among the trout. Even so, I'm still hoping that this downstream stretch will gradually prove more attractive.

Having done my best to put you off, I'm sure that some of the more foolhardy among you will still find a spot where you simply 'must' build a weir. To tell the truth, I've just come back from the river, from the stretch with my little weir, and a very pleasant feature it makes. It has added character to a rather monotonous bit of river and the sound of running water enhances its attraction. The sight of a few trout rising is a bonus, so I suppose I must really be 'pro-weir', provided that care is taken to select a good site.

Permanent Weirs

In this section, I'll deal initially with the construction of one which you are intending to be a permanent feature, and give some suggestions for more temporary weirs later. The materials, as for groynes, are angle irons, wire netting and poles – really strong poles, especially for the sill of the weir. One extra item for weir-building is plastic sacks, lots of them! You should have no trouble getting hold of these; your friendly farmer will be able to supply all you need. Their disposal may be one of the farm foreman's little perks, so it may be advisable to make sure that he is not out of pocket on the deal; you need allies, not enemies. While on the subject of plastic sacks, a couple of tips may not be out of place. Only half-fill a sack, as it will then be much easier to handle, especially if you are a professional weakling like me, and the sack will also be much easier to shape into a nice block. It's not really necessary to tie the sack up, merely fold the empty end underneath with the fold facing upstream, though in actual fact I do tie most of mine, as then you don't get quite as wet putting them in place. One last point, make a few holes in the sack with a garden fork to let the air out – you don't want your sacks to float away down the river.

The actual construction begins with making a double groyne, though this time the function of the groyne is to protect the bank from erosion and to try to prevent the water from finding a way round the side of the weir. These groynes should be reinforced with some of the plastic sacks, and the bankside end should be appreciably higher than the expected new water level. The end out in the river should be just higher than the sill of the weir.

The width of the weir in relation to the total width of the river clearly depends on what effect you are seeking to create downstream. If the flow is not too sluggish and your only aim is to raise the level of the water upstream, then the weir should be almost the full width of the river. The two groynes will need to be at a very shallow angle indeed, and with this type of weir there will be fewer problems with scouring and back eddies.

If, however, the river is very sluggish and uninteresting you may wish to combine some of the effects of a groyne and a weir, producing something more like a small mill pool. In this case, the pair of groynes

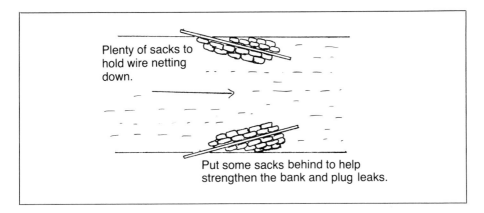

Plenty of sacks to
hold wire netting
down.

Put some sacks behind to help
strengthen the bank and plug leaks.

Fig 24   Initial stage of building a weir.

will need to extend much further out into the river. With a weir like this
there will be much more scouring – enough to undermine the angle
irons' stability, which therefore need to be long and strong. More plastic
sacks will be needed to ensure that the groynes are strong and
watertight.

Before any attempt is made to raise the water level, it is best to leave
this initial work for several months so that the wire netting can clog up.
If you are lucky enough to have some floods during this period, even the
netting where the new water level will be may become moderately
blocked. The plastic sacks will gradually settle into place, and some extra
ones may be needed to plug gaps. Finally, all the softer silt, sand and

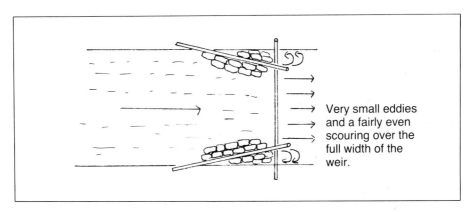

Very small eddies
and a fairly even
scouring over the
full width of the
weir.

*Fig 25   Angle of groynes where weir is only used to raise
the water level.*

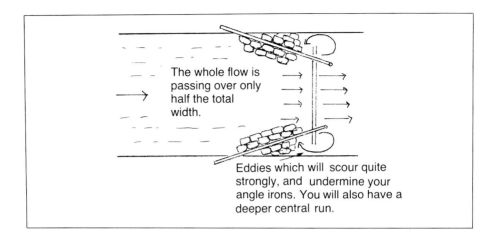

The whole flow is passing over only half the total width.

Eddies which will scour quite strongly, and undermine your angle irons. You will also have a deeper central run.

Fig 26   Narrower weir to raise water level and provide more scouring.

mud will be scoured out, leaving a fairly hard river bed where the sill of the weir will be. The sill itself may, in point of fact, have been put in place in this initial work, but the water is still free to flow under it.

Once the groynes and the river bank seem stable it is time to complete the work. Basically, all that needs to be done is to staple wire netting to the sill of the weir, leaving the netting long enough to extend several metres upstream on the bed of the river. The whole lot then needs to be held firmly in place with plenty of filled plastic sacks. Uprights should be placed along the sill to stop the netting bulging out, and to give extra strength.

When building the weir, do not stint yourself of materials. There is a lot of hard work involved, and it is all wasted if the structure collapses within a few weeks. There should be a pile of plastic sacks already filled with earth on hand. The uprights used to stop the wire netting bulging out along the sill of the weir should be strong and go well down into the river bed. The wire netting – I suggest 25mm mesh for this job – should lie on the river bed for at least two metres (7 feet) upstream of the weir and be completely covered with the plastic sacks. In this way, it will be held firmly in place and much erosion will be prevented.

If you are lucky enough to have a straight tree trunk three or four metres (10–15 feet) longer than the river is wide, you will be able to dig the ends into the bank, so that there will be a very strong sill for the weir. This is what I have done in the weir which I am in the process of building. If, however, the sill of the weir is to be supported by angle irons, these angles need to be very strong and deeply embedded, as they

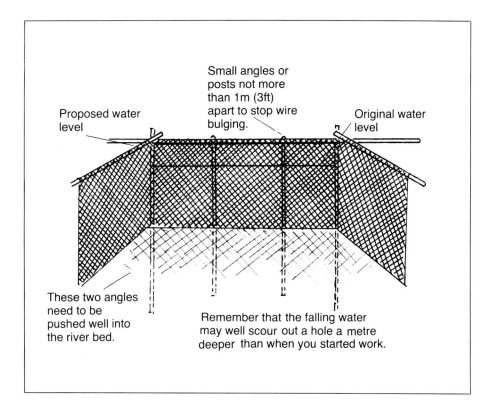

Small angles or posts not more than 1m (3ft) apart to stop wire bulging.

Proposed water level

Original water level

These two angles need to be pushed well into the river bed.

Remember that the falling water may well scour out a hole a metre deeper than when you started work.

Fig 27   Basic construction of weir.

take a very great strain. I suffered one minor collapse and was lucky not to lose the lot, so be warned!

Finally, let me stress once more the importance of selecting your site carefully. The weir needs to be on a straight, where raising the water level will give real benefit by increasing the amount of holding water. The banks on both sides of the river need to be firm and stable. I remember one incident a couple of years ago when the club concerned (not mine!) was putting in a grid for a fish stop. One side was really only a silt bank and the site chosen was on a bit of a bend. The bailiff was slack about keeping the grid clear, so the grid tried to turn into a weir. The water had other ideas and just scoured a channel back through the silt bank! Try to ensure that both the river bed and banks are firm at the point you wish to build your weir, as there will be quite powerful scouring, even on good gravel. If the river bed is too soft, your work is much more likely to be undermined and the whole lot will collapse.

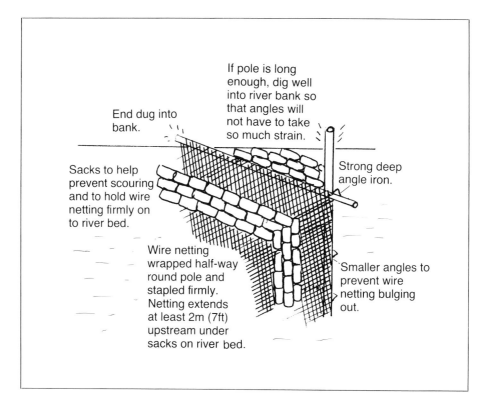

If pole is long enough, dig well into river bank so that angles will not have to take so much strain.

End dug into bank.

Sacks to help prevent scouring and to hold wire netting firmly on to river bed.

Strong deep angle iron.

Wire netting wrapped half-way round pole and stapled firmly. Netting extends at least 2m (7ft) upstream under sacks on river bed.

Smaller angles to prevent wire netting bulging out.

Fig 28 Detail showing one side of weir and the reinforcement needed, seen from above.

Temporary Weirs

I have no direct experience of more temporary structures, not, at least, of their construction. Our local farmers inadvertently produce very low weirs for us from time to time when their barbed wire, stretched across the river to keep the cows in place, collapses. Weeds, branches and rubbish collect with great rapidity and quite often make a little weir, raising the level upstream by several centimetres and scouring the bed of the river clean for a surprising distance downstream. Fish find this quite attractive as the rubbish provides good cover. Certainly, on a little river, three or four strands of barbed wire stretching from bank to bank will often make a low weir. The ends need to be firmly staked and only the top strand of wire needs to be just above the water. The advantages are simplicity and cheapness; they take only a few minutes to make, and

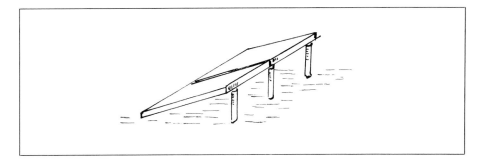

Fig 29   A temporary weir made from paving slabs.

even less time to dismantle. The primary snag, however, is that there is no guarantee of success; luck is bound to play a part in the proceedings, and there is also the aesthetic objection that the accumulation of rubbish against the wires looks unsightly. Even so, such places provide a good hiding place for the trout and an extra habitat for the insect population.

I have also seen illustrations of temporary weirs made by resting paving slabs on a line of posts which have been driven into the bed of the river. I have no personal experience of this sort of structure, but it looks as though it should make an effective low weir. Clearly, this will only work in a shallow stream with a reasonably level bed. The idea is that the slabs can be taken out in the winter, allowing the water to flow at its normal level. On such small waters, of course, the simplest of all little weirs can be made by using half-filled plastic sacks, though even here a length of wire netting will help to keep everything in place.

The New Weir: Stop Press!

As I said earlier in this chapter, I am in the process of building a new weir; what a saga! Fig 30 shows the tree trunk I have used in position to form the sill of the weir; both ends are dug well into the bank. This really is a most satisfactory way of starting the job, as it is very much stronger than hammering uprights into the river bed to take the weight of the weir. The two groynes are already in position, and I am busy putting sacks of earth in place to protect the banks from erosion. All this proved a pleasant summer occupation – so far, so good!

I only needed to wear thigh waders to do all this initial work. I then waited, just as I have suggested that you should do, for floods to scour away the loose material on the river bed. I could never have foreseen the great gale which struck Britain in the autumn of 1987, and the floods that came with it. A huge amount of debris and branches piled up against the sill for the new weir and it looked more like a beaver dam, except that the water still found its way underneath. The current had certainly scoured the river bed with a vengeance. I could just manage to work to clear all the rubbish wearing breast waders, but even then I came dangerously near the Plimsoll line! When the time came to make the actual weir, there was no question of just folding the sacks to keep the earth in place; in that deep water they were all securely tied up, and I still managed to get wet.

Having seen the amount of scouring which had taken place, I used posts a full two metres (7 feet) long, hammered right down into the river bed to hold the wire netting in place along the sill of the weir. I reckon that it is just as well I did as, within a few weeks of completion, a weir pool almost two metres (7 feet) deep had been scoured out, throwing up a great bed of shingle six or seven metres (22–26 feet) further downstream, with another quite deep pool forming downstream of that.

My troubles weren't entirely at an end. We completed the weir just

Fig 30   Sill for the new weir: a strong willow trunk has been dug into both banks. I am putting sacks in place by the groynes to protect the bank from erosion.

Fig 31   The new weir and an ever-expanding weir pool.

before Christmas and I was hoping for a spell of crisp, seasonable weather to allow the weir to settle into place. No such luck, the ground was still saturated from the autumn floods and every time it rained, which it did with monotonous regularity, the river rose in flood yet again. Quite a lot of reinforcement work was needed around the groynes, especially on the far bank, as the water seemed very keen to find a way round. This was partly my fault for not doing as I've told you to do, because the river is not absolutely straight upstream of the weir, which greatly increases the risk of erosion. This problem was made worse because the banks are very soft. We had to put in quite a lot more sacks to protect the bank, but now all seems well, as Fig 31 shows. Actually, I have increased the depth upstream of the weir a few centimetres more than I had intended, but I hope it will all be to the liking of the trout. This stretch certainly fished much better than usual during the 1988 season, so the signs are promising. In fact, a wild brown trout of 4lb 1oz has just been taken from it, almost certainly a record fish for the water.

NARROWING THE RIVER

The construction of groynes and weirs seems to be the most direct way of improving both the state of the river bed and the fish holding qualities of a stretch of river. However, as I have said earlier, one of the problems with lowland rivers is that, for much of the year, there is not a sufficient volume of water coming down; abstraction will almost certainly have taken its toll and dredging may well have made the situation worse. The net result is that there are now many stretches of river far wider than they need be, relative to the volume of water which is flowing down them most of the time. If the water on one side of the river is sufficiently shallow it is possible to make the river narrower by making a new low bank – a berm is the word for it.

A bank like this will always be lower than the main bank of the river, so that flood water will flow quickly over the top, but in times of normal flow all the water will be confined to the narrower channel, giving a rather swifter current and helping to keep the bed silt-free. I have done work like this quite successfully in four different places along one stretch of river, but the key to this sort of work is the availability of suitable materials, and the ability to move them to where they are needed on the river bank. Through the kind help of a friend I was lucky enough to get a lorry load of wooden pallets just for the price of the transport. Added to this, our river flows past a little wood called Osier Carr so, as the name implies, I have an almost unlimited supply of willow stakes and brushwood to help hold everything in place.

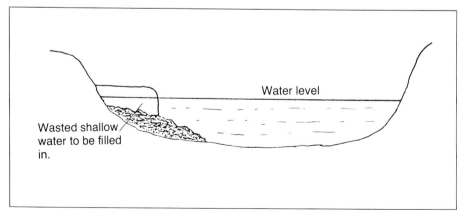

Fig 32 Suitable site for a berm, or low bank.

Fig 33   *The river has been narrowed upstream as far as the*
big willow. The pallets have disappeared under new
vegetation.

This first river-narrowing job that I tackled was also the biggest. There
was a straight 30–40m (33–44yds) long, one bank of which was little
more than a shallow swamp. This is the point at which you need all
hands on deck, and a bit of help from your landlord as well. The
gamekeeper borrowed a tractor and trailer from the farm, and club
members helped to load the pallets, so the whole lot could be brought
almost to the work site.

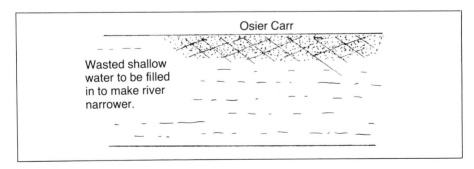

Fig 34   *Length of river to be narrowed.*

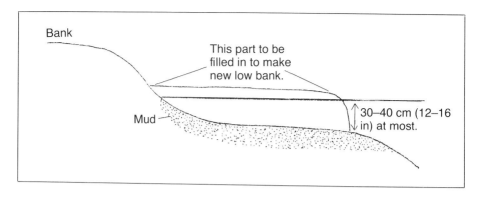

Fig 35　Side view of a berm.

I doubt, unless you have a very large amount of material available and a labour force to match, that you will work very satisfactorily once the depth of the water is much in excess of 30cm (12in). As I have just said, the work is messy so dress accordingly and be prepared for immersion. Start at the upstream end of the stretch which you intend to narrow and build yourself a base gradually extending as far out into the river as you intend to go.

By dropping a few bundles of brushwood on to the silt first it is possible to lay the pallets reasonably level; the whole job will look terrible if they are left sticking up at odd angles. Remember, too, that until they are waterlogged the pallets will float, so things need to be held securely in place. This is where the lavish supply of willow stakes and branches is so useful; tying wire, binder twine, hammer and staples will do the rest.

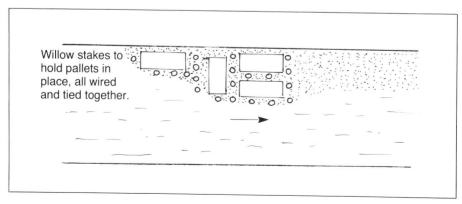

Fig 36　Method of narrowing river using wooden pallets.

44

Once the base has been made, pull up all the mud you can reach with your chrome or similar implement and deposit it on top of the pallets, so that there is now a genuine extension to the bank. From now on, work downstream in the same way, extending as far out into the river as you wish to go or the depth permits, staking and wiring all the pallets very firmly before pulling more mud on top of them. As shown in Fig 36, do not have the pallets too close to each other. Apart from allowing vegetation to grow up more easily and making the whole bank more permanent, a greater area will be covered with fewer pallets (and there will be a very nice trap for the unwary, who step into the gap and get very muddy indeed!). Do not try to tackle too great a length in one go; it is hard work, especially pulling up the mud, and the more mud that is pulled up and put on top of the pallets, the more stable the new bank will be.

Let me confess straight away that our first effort was not a pretty sight, even though it was very effective. Nature is doing its best and our rather untidy work has now almost completely disappeared under the new vegetation. Subsequent efforts have looked better much more quickly. One refinement is to try to make a clearly defined river bank. One of our number, less ham-fisted than the rest of us, built an edge to our work by pushing willow stakes into the silt and weaving a bank from thin willows and brushwood, rather like crude basketwork. Lots of this has grown, and provides a bit of shade as well as a new river bank. The best-done job just does not show; this bank has become firm and completely grassed over. The only evidence of what is beneath is provided by a few posts which were used to hold the pallets in place; they are still sticking up and ought to be sawn off, when I can find the time.

This type of work does need a labour force of four or five of you, if it is not to be too much of a chore. If machinery is available, jobs like this are much simpler. A couple of years ago the water authority was replacing a footbridge which had been damaged. It was a shame to see expensive machinery standing idle while bits of the preparation work were being done. We managed to get a little free dredging done, and we also fenced off one muddy bay, using willow stakes and wire netting. The operator kindly dropped the silt from the main channel in behind the wire netting, doing in a few minutes what would have taken us several days' hard labour, and clearing the main channel much better than we could possibly have done. Some of the willow stakes which were pushed in to hold the netting in place are now growing well, and the spot

Fig 37 *Jim's basket-work bank, some of which is now growing quite well.*

is beginning to make a pleasant feature on what is a fairly treeless stretch. The area filled in is still too muddy for cows to walk on, so the trees can grow unmolested.

A little job like the one just mentioned is very much a case of taking an opportunity when it arose. You won't often find an expensive piece of machinery standing by your river, with an operator willing and eager to

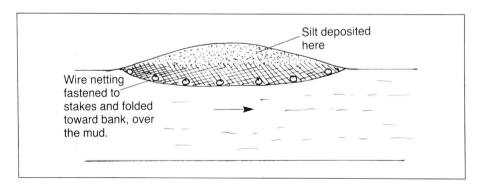

Silt deposited here

Wire netting fastened to stakes and folded toward bank, over the mud.

Fig 38 *One method of narrowing when machinery is available.*

do some work for you. However, if you have a considerable amount of water which could be improved in this way and you are not lucky enough to have the amount of material I had for some of my river-narrowing jobs, it may pay you to hire a machine and have some work done. However, apart from seeking permission from both landlord and river authority, the ease of access and the firmness of the main river bank must be taken into consideration when deciding whether or not to hire. A job is simply and speedily done if a machine can work directly from the bank; if the machine has to stand on a mattress because the ground is too soft, the same work takes very much longer. Should you tackle this kind of work using machinery, all the preparation work must be done before. Willow stakes and wire netting should all be in place, so that the machine operator has only to drop the spoil in behind and move on. If you do hire, get a precise tender beforehand, so that you know just how much money you are going to spend. Better still, make sure that the contractor is a member of the club!

The topic of machinery and dredging is a vexed one, but don't be put off by some of the appalling work which was done in the past and, alas,

Fig 39   A river narrowed with poles and wire netting. The dredger has dropped spoil in behind.

is still being done in some places today. Properly planned, and with a skilled operator in the machine, dredging can do a super job. One stretch of river which I look after had become so badly overgrown that, in summer, it was almost impossible to put a fly on the water. The banks, too, were very boggy indeed in places, so soft in fact that a couple of summers ago I went straight through the top crust and sank to just above my thigh waders in liquid ooze. Fortunately it was a warm day and my clothes, which I had to wash, soon dried out.

The river, as I say, was in a poor state and even the water authority, whilst pleading poverty and financial constraint, agreed that something must be done. We took care not to canalise this pretty stretch of stream; narrow necks were left quite deliberately in several places, so that we would keep sections of fast current, and nowhere was the river widened to such an extent that there would be any danger of it silting up again immediately. The machine arrived and I explained to the operator what I wanted, above all that the gravel from the bed of the river should not end up on the bank. Scarcely a single pebble appeared with the spoil, the boggiest bits of bank were built up and smoothed off, so that, once the mud had dried out, it was good firm walking. Every so often a trout was excavated; Dennis would pick it up gently in the bucket and drop it back into the river. Finally, we would march off together on a crayfish patrol and put the ones we found ambling about on the mud back into the river. That stretch of nearly 200m (220yds) is now a delightful piece of water, the banks are all nicely grown up again, and it is a very pleasant bit of fishing.

That was the second job that the water authority did for us in the last few years. 'For us' is a bit of a red herring, as such large sums of money are not going to be spent by the engineers just to improve the fishing. A little pressure from the farmers, complaining that their drainage is being badly affected, is often what is needed. The first job the water authority did for us was to tidy up the straight, upstream of a mill. This is a canal-like stretch of water but, here again, boggy banks were transformed into firm walking and the bed of the river was left intact. On both occasions, we wrote to the local division and to the water authority headquarters to say what a super job had been done. So, don't be too afraid of the dredger. Incidentally, where work is to be done, the authorities are under some sort of obligation to maintain and improve fisheries. Much of the damage which has been done is the result of ignorance, not malice. A good working relationship with the engineer responsible for your stretch of river will pay handsome dividends.

You are much more likely to want to make parts of your river narrower rather than wider. The more the speed of the current diminishes, the more readily silt is deposited, and too much silt is no good for the trout *or* their food supply. The late Frank Sawyer wrote much about the beneficial effects of chalk in helping to break down silt and in improving the quality, the palatability even, of the food supply. We have certainly put quantities of chalk into our water in the past and more recently have had the river treated with Nautex, the market name for a very fine type of chalk mined in the Champagne region of France. The theory is that the very fine particles of chalk penetrate the silt, allowing oxygen to enter, and bacteria can then work at breaking down the silt. Provided that you are dealing with organic silt, dead weeds and the like, it seems that this treatment works, but I find the results hard to quantify, especially as it was used in conjunction with so many other river improvements. You ought to be able to obtain impartial advice from the water authority and I will return to the subject when discussing lake fisheries.

STONES AND OBSTRUCTIONS

Groynes make fairly large-scale obstructions, but any observant fisherman cannot fail to notice how often fish are to be found near much smaller obstructions. Old push bikes, prams and tin cans create a bit of turbulence, which fish seem to find very attractive. I'm not suggesting that you raid a scrapyard and drop your ill-gotten gains into the river, simply that the fish holding qualities of a shallow stretch of the river can be increased quite considerably with very little effort. Big flints or lumps of concrete create quite a lot of turbulence, besides providing extra habitat for fish food; a trout will often lie, ready to intercept its food, immediately upstream of a stone, and may well find satisfactory shelter in the little hollow scoured out downstream of the obstruction. We have done this on a couple of shallows and find it moderately useful, and flints and lumps of concrete soon settle in and are not unpleasing to the eye. Once the water is deeper, it is hard to assess any benefits and the obstructions may silt up. Provided that the materials can be easily transported to where they are needed, it is a little improvement well worth making.

That concludes what I have to say about work on the river channel itself and what I would call the construction side of the job. I'm not suggesting that your stretch of river will need all these different forms of treatment, but it is important to have a clear idea of what improvements are needed and to work out a plan of campaign. Observation will show what features seem to be attractive to the fish and the aim should be to provide as many of these features as possible. I'm certainly not suggesting that my way is best. We are quite convinced that the improvements we have made have been a great success, but there are always better ways of doing a job. It is only in the field of groyne manufacture that I feel confident that my way *is* very good, both on the grounds of cost and simplicity. Much of the work you undertake will depend on the materials that are most readily available and the kind of help your group is best able to provide. It is surprising what a variety of skills half a dozen individuals can bring to your working parties. I haven't mentioned money. If you are very rich, you can employ people to do the work for you, but you will miss quite a lot of fun and not have the same sense of achievement. Remember, don't expect results overnight, and finally, may you have more successes than failures.

3
Water Plants and Bankside Cover

Chapter 2 dealt with the condition of the river bed itself, and possible action that could be taken to improve it for the benefit of both the fish and their food. I would liken that part of the job to building a house, and now that the house is built the question of the decorations and the furniture arises, the trees and bushes growing on the banks being the decorations, and the furniture, the weedgrowth in the river. All of them need planning and maintenance and, if this is done successfully, the 'home' will be even more attractive to the fish. A healthy lowland river will support a luxuriant weedgrowth, to such an extent that the watercourse will be completely choked by high summer, unless some weed cutting is done. Some water plants are very attractive indeed. Ranunculus, the water buttercup, springs to mind, with its carpet of dainty flowers. It is easily my favourite water plant, lovely to look at, easy to cut and with a root system that does not trap too much mud. Water parsnip, too, is very attractive and provides a good habitat for fish food. Both these plants prefer the faster-flowing stretches and are well worth encouraging. Other water plants have less charm and definitely come in the weed category. Ribbon weed is one such, and blanket weed, though it harbours lots of insects, is the dreaded end. The last-named flourishes in the slow stretches, especially in over-enriched waters. It traps all the silt that floats down and forms a slimy layer of ooze on top of the river bed, so dense sometimes that even the winter floods fail to shift it and the mud problem starts up all over again. It adds to its list of sins by choking other more desirable plants. Some of the other plants more commonly encountered are starwort milfoil, potamogeton and Canadian pond weed. Starwort is a very pretty shade of green, but it does tend to trap a lot of silt, as does the Canadian pond weed, so both need to be well controlled.

Clearly the plants most numerous in your stretch of river will be those most suited to it so, however hard you may try, you cannot really change the natural fauna. Even so, the improvements that will have been made in the river channel, and the speeding up of the current in many places, will have modified the regime a bit and should enable some of the more desirable plants to flourish, though it may be necessary to import them from elsewhere initially. I personally found that clearing the silt also cleared out lots of starwort and pond weed. Regeneration of the more useful plants started very well, but has been inhibited more recently by an excessive swan population early in the year. The wretched birds took a delight in uprooting everything. The swans gradually departed and then the blanket weed grew and did its best to choke the new plants. Latterly, and worst of all, excessive suspended solids, coming from iron deposits exposed when an internal drainage board employed a 'cowboy' to dredge a small tributary, cloud the water even in times of low flow. Fortunately this pollution is not toxic to the fish, but the reduced light penetration has had a devastating effect on weed growth. You may not suffer this particular problem but there will be others; the path of true love never runs smoothly. The important thing is never to give up (I keep telling myself!).

WEED CUTTING

Weed is vital to the river, helping to provide oxygen, and giving shelter to fish and insects alike. However, a plan must be formed to encourage the good and root out the bad. The major task is to organise the weed cutting. The need for this varies to a very great extent from one region to another, and even in different rivers of the same region. In Hampshire and Wiltshire, weeds have often been cut once before we see any at all in our Norfolk rivers. In some rivers, the weed needs to be cut at least twice a year, whilst even before my pollution problems, I could almost always get away with only one cut, in about mid-July. The biggest problem of all is who cuts the weed and, far worse, who gets rid of it? Our region is almost all arable farming, the farmers want the land drained, and weeds cause flooding. Before they are cut, weeds can add half a metre or more to the height of the river, as they hold the water up so much. As weed cutting is a vital drainage job, the water authority undertakes the work, cutting until the river becomes too shallow for their boat to negotiate. There may be inconveniences and the job may not be done exactly as you might wish, but just be profoundly thankful if someone else cuts the weed for you! I almost always sit in the boat with the men, when they cut our stretch of river, just to temper their enthusiasm; rather like an operator doing some dredging always wants to dig out a bit more, so the weed cutters always want to cut a bit more. It is vital that a reasonable fringe of weed is left, but it must be remembered that the authority is cutting the weed for drainage purposes, so that the water can flow freely and the levels stay down. The farmer, who has to pay a drainage rate, does not want his meadows under water, so a compromise must be reached. A few years ago the weed cutting was left until far too late in the season, despite many protests; this also happened to be a season when the growth was remarkable. One farmer couldn't even drive his tractor over what was normally a hay field, and water levels had risen so much that the river even flowed over the *top* of some bridges! So remember, the weeds are being cut for the farmers, not to improve the trout fishing. It is, however, possible to liaise with those who are doing the work and, as far as I am concerned, they are very co-operative. In the end you will hopefully have a river in which a good clear channel has been cut, but still with plenty of weed left to provide cover for the trout.

When I know that the weed cutting boat is on its way, I then cut the narrower stretch by hand. I do a little using a scythe, but this is quite slow work. Obviously, if time and labour are unlimited, cutting like this is ideal. The river can be tended like a garden, banks of weed being left to deflect the current in the most useful direction; a fly can then be cast into the open spaces and the trout will still have plenty of places to escape from the angler. Life is not ideal, however, and as time and labour are by no means unlimited I find a chain scythe much the best implement to use. The water authority has kindly made up a small chain scythe for me, with only about half a dozen blades in it. This is not the back-breaking implement that the old chain scythe was, but the blades are still heavy enough to keep quite close to the river bed as you work. A friend and I put on chest waders and we can work quite quickly and, with such a relatively small cutting edge – about 2.5 metres (3 yards) in all – we can cut the weeds quite selectively. We always start at the downstream end and work up, so that we are always cutting into clear water and can see exactly what we are doing. The weeds float away down the cleared channel. Working like this, it is amazing how rapidly the water level is lowered and how fast the weeds float away. We never leave the river looking bare, and try to strike a balance between drainage and fishery interests.

Fig 40   Not our sort of river. Work here would be like giving a strawberry to an elephant.

Our club possesses a very light chain scythe called the 'Hutton Clear Weed'. This is not much use to us, as the blades, which are rather like the teeth of a saw, quickly clog up with blanket weed and won't cut anything. However, another club has borrowed it from us and they find it very useful on their weed growth. We certainly need the greater weight and sharper cutting edge of the old-fashioned knives for our work.

The law says that cut weed must be removed from the river and not allowed to float downstream to annoy other people. On some rivers, to try to minimise the disturbance, weeds are cut at the same time on several different fisheries. We are lucky to have an authority which does all the weed cutting and pulls the weed out of the river with machinery. All we need do is to make our cutting coincide with theirs, so that all the weed floats off downstream at the same time, and the authority pulls the whole lot out. If you are not lucky enough to have some such arrangement, I feel sorry for you – any fool can cut weed, but it needs a whole team of lunatics to pull it out, and strong lunatics into the bargain; weed is very heavy indeed. This is the one job which I suggest you put out to contract, if at all possible, and consider it money well spent. Even then, somewhere has to be found to put the cut weed; it can be a fairly unsightly, fly-covered mess if left. During the cutting, a boom (a couple of ropes or a length of netting) must be stretched across the river to catch the weed as it floats down. This needs to be securely anchored, as it must bear quite a heavy load at times. In any case, never cut too much weed at once; do a section and pull out the cut weed before starting to cut another lot.

I mentioned the weed-cutting and clearing propensities of swans earlier in this chapter. Some years, thirty or forty of them descend on one of our fisheries in the spring, annoying the fishermen and pulling up the weeds. I used to get somewhat hot under the collar about this but when, later in the season, the weed boat had not arrived and anglers on other stretches were hard put to find somewhere to fish, I began to have a soft spot, even for swans! My members were all fishing happily, and the swans had mostly flown away. I realise this is not an ideal solution to the weed problem but, as we rely heavily on stocked fish for our sport, the actual damage done by the swans is limited.

In the early stages of your project, you will almost certainly want to drag out unwanted weed, mud, roots and all. This is back-breaking work. Also, some of the weeds act as massive silt traps and great banks may have built up over the years. Such banks can be cleared by using groynes to deflect the water and wash them away. Sometimes, however, it is useful to pull the mud out with a chrome and use it to build up low-lying and boggy banks, but it is important not to tackle too much in a session, unless several of you are taking turns with the chrome. I have been pleasantly surprised by the improvements made to some stretches; banks have become firm, and the river channel itself has not silted up again, probably because of all the other improvements which have been made.

Before turning to more constructive matters like planting, I should mention one other weed which needs to come under constant attack and should be cut at least twice a year. This villain, attractive enough in its proper place, is the common reed (phragmites) which grows very well on swampy ground and gradually marches out into the river, trapping silt as it goes. It builds up a big bank, deflects the course of the river and erodes the far bank. On dry land the reeds are easily cut with a strimmer but, out in the water, I always cut them with the slasher (*see* Fig 6, page 18), reaching as deep down as I possibly can. This often means that I am cutting right down into the mud and do a very effective job. The reeds don't much like being cut regularly and, within a few years, the numbers growing will have been drastically reduced, and in some places they will have been eliminated. In the early days, we also set about them with the chrome, pulling them out by the roots (rhizomes). On one occasion, a tree came down in a gale and diverted the current right through a reed bed. The fast flow cleared all the silt and the rhizomes were laid bare. It was the easiest of jobs to chrome them all out. There's hardly a reed left there now, another example of how the water can do the work for you. One keeper I know had a nasty reed bed and he gave it a short sharp dose of Round Up. I'm no chemical gardener myself, because I am never quite sure of possible side effects, but his results were highly satisfactory and as he looks after one of the best fisheries in the country, who am I to argue? Whatever methods you adopt, war must be waged against the common reed, otherwise the river channel will become choked.

TRANSPLANTING

It may be necessary to move some weed from more favoured parts and transplant it to barer areas, or even to import some from another river. The practice is similar to weed cutting; it's very easy to dig up some weed, but it takes a lot longer to replant it. Use plastic sacks to carry weed from place to place. I have tried various ways of planting weed;

putting roots into fibre pots and sinking them into the river bed, or putting a bundle of roots into old sprout or onion nets and weighting them down with a few bricks, but I have always had the best results when I have planted the roots directly into the river bed. This means that you can only work in shallow water, certainly less than a metre deep. The deeper the water, the wetter you get, so it is always advisable to pick a warm summer's day for weed-planting operations. Don't waste time planting too much at once; if the plants like their new home, they will flourish and spread. Don't plant everything in one spot; try several well-separated sites, and there will be more chance of finding one that really suits the plants.

PLANTING

Growing trees and bushes is a much easier, and drier, job. As I mentioned earlier, fish like a roof over their heads, and the water plants are not the only things to provide this roof. Bridges, for example, have an unfailing attraction for fish and fishermen alike; there are almost always at least half a dozen trout under the road bridge at the top end of one of our stretches of river. Any overhead cover will do. A couple of years ago, our punt had been moored in one place for quite a long time and a few trout were actually seen to be living underneath it; sometimes they came out in the open to feed, sometimes they could be seen brushing along the sides of the punt, presumably dislodging insects that were living on the sides.

Considerable use can be made of trees and bushes to provide roof, shade and even a few droppings of food. Eventually some roots will grow out into the river and these provide super cover for fish, though if there is too great a mass of roots, they may form a strong back eddy and erode the bank. There has to be moderation in all things, however. The river cannot be allowed to disappear under a complete screen of trees; apart from making fishing almost impossible, no water plants would grow in the gloom (in fact shade is very useful for inhibiting weed growth). I have one very pretty stretch of river which flows through a fairly thick wood and it is nothing like as productive as the stretches upstream and downstream of it, where there are less trees. It makes for very pleasant fishing to have one bank open and the other side bushed. It is better to have the bankside cover on the side of the river from which the trees will give maximum shade from the afternoon sun. This may sound simple but there are snags, most of which have four legs – sheep and cows seem to think that young trees are planted just to vary their diet! Your plan of campaign will also be governed by the farmer's wishes and the type of agriculture practised. There are very often restrictions imposed by water authorities on planting trees too close to the river bank, because there is the chance they will get in the way of dredging machinery. These restrictions do not usually affect me, as I don't allow my trees to grow very large; I am much more keen to have some large, overhanging bushes.

The easiest trees to grow are willows and alders, both of which flourish in damp ground. Any willow branch pushed into the earth is almost sure to grow. Alders are supposed to do so as well, but not for me unfortunately! I have always had to plant alder saplings, but they are usually quite easy to find. As you must have noticed, there are many varieties of willow ranging from those which grow into big trees to others which are much more bushy. I have a line of crack willows along

Fig 41   Two newly-pollarded willows on the left and right. The middle one was done the previous year and shows how quickly the trees grow again.

one bank of my river, planted before my time. They are just about the commonest of all the willows, but I don't think they are much use for providing cover, though they are nice enough to look at. Unfortunately, they need pollarding every few years, otherwise the branches will grow too big and heavy and, in the end, the trunk will probably split and the tree will die. I always plant a few pussy willows, if only to delight in their lovely yellow catkins. Hawthorn, blackthorn, sloes and elders all play their part near the river bank, though I haven't needed to plant any of these. I merely do battle with them from time to time with saw and secateurs, wearing the thickest of gloves when fighting with the thorny ones! Really, I suppose, we come back to what I have been stressing all along, the question of variety, so that there are different shades, flowers and berries to enhance the beauty of the river bank. The dull, sterile monotony of conifer plantations supplies all the evidence that is needed in favour of variety.

Tree planting falls into two categories, one to provide overhanging shade and cover as quickly as possible, the other a very much more long-term project, to provide the trees which will be a visual amenity in years to come. This second job is usually done in close co-operation with the owner, whose idea it may be in the first place. He may well wish to improve the appearance of his property, and quite a lot of preparation is needed for this sort of work. If there are sheep and cattle about, the area to be planted must be securely fenced off. You could wire off a substantial length of bank, well back from the water's edge so that trees can be planted on firm, well-drained land, out of reach of all but the highest floods. Even then, the young trees need to be properly staked and provided with plastic guards to thwart the rabbits.

For immediate results, willow and alder are more useful and these should be planted as close to the water as possible. Don't make one of my mistakes, however, and push the willow stakes in too close to normal water level, in the hope of getting more shade more quickly. Once the stakes have grown into little bushes, even a small flood will half cover them and pile up a lot of debris against them. The pressure is then quite likely to uproot them and take a chunk of bank with it, the last thing you want to happen. The willows and alders should be planted in firm ground, on top of the bank and about a metre (3 feet) back from the edge. In this way, they should be relatively safe from floods, and the growing roots will help to stabilise the bank. Put in far more than you need as it takes several years to grow a decent bush and only seconds to cut one down. Unless you want a fully-fledged tree, prune them once a

*Fig 42   This willow was planted much too near the water.
The bushes provide good cover at the moment, but floods
must soon sweep them away.*

Fig 43   These trees are better placed, but still too close to the water. The ground is very soft and floods have already pushed some of them over. They will still grow as bushes and help to consolidate the bank.

year. I always cut the trunk off at about head height, so that the trees are easy to manage and unlikely to be blown down in a gale. Regular pruning ensures that the trees grow very bushy. It is helpful to drift down the river in the boat most years, not only to cut back the bushes a bit, but also to make sure that all the lower branches are well clear of the water. This leaves some room for the skilful to cast underneath, and gives flood water less chance of causing damage.

Quite a lot of work has been covered in this chapter, but let me emphasise once again that you can't do everything at once. After a few years, observation will soon show which areas are favoured by the fish and which ones need to be made more attractive. In the suggestions I have been making, I have assumed that I am writing for people who are not going to spend every moment on the river; most of us have wives and families, and other interests apart from fishing, so for us keepering is not a full-time job. Many of the jobs that have been done will bring permanent benefit to the river, benefits which should increase in the course of time as the river bed is scoured cleaner and the new trees and bushes provide better cover. Maintenance will become increasingly important, to ensure that the best holding spots remain attractive to the fish and accessible to the fisherman.

4

Stillwater Fisheries

I was recently reading through the Fishery Reports for 1986, which are sent to me by the water authority, and as always the section on fish mortalities caught my eye. Why is it that death, doom, disease and disaster unfailingly hit the headlines in preference to more cheerful and constructive items? The lakes of the region, it seemed, were littered with dead and dying trout. Water quality problems inducing stress, and stress caused by high water temperatures were two of the reasons advanced. On several waters argulus infection was diagnosed, elsewhere ligula reared its ugly head. Add to these woes a few outbreaks of bacterial infection and a couple of doses of proliferative kidney disease and you should be starting to get the idea. Of course, I wouldn't be wanting to put you off starting a stillwater fishery, though I do remember one year, on our lake, when we all ended up coarse fishing from the end of July onwards. The few trout left alive after a disastrous infestation of eye fluke couldn't see anyway and there was no point in wasting the fishing completely. Not all the disasters of 1986 occurred in waters controlled by rank amateurs, some were in well-established commercial fisheries. As I have just said, I wouldn't want to put you off the idea, but I reckon that looking after the river is a sort of rest cure when compared to the hazards of running a lake fishery.

There is certainly a demand for trout fishing in smaller still waters. Some anglers definitely want easier fishing than that offered by the big reservoirs. Many find fishing these vast waters a daunting prospect, and when the trout aren't rising there is always the feeling that the nearest fish is at least a mile away. I know the feeling only too well. Many years ago, we were fishing Durleigh Reservoir on a cold day in early spring; not an offer had come our way, not a rise was to be seen. 'Albert', we said to the keeper, 'there are no fish in your reservoir!' Albert smiled, said nothing, and threw a handful of pellets round the boat; at that, the water boiled!

Unless you are the eternal optimist, bank fishing from the reservoirs, as opposed to boat fishing, can be even more depressing, especially when no fish can be seen moving; certainly, the trout are not evenly distributed and to go from one shore to another is a major expedition for the fisherman! Only recently, we were boat fishing at Blagdon Reservoir and doing every bit as badly as the bank fishers on the south shore. No fish were showing anywhere and, in common with most of the boats, we only had a couple of fish. The anglers fishing the north shore had no complaints and were catching fry-feeders quite steadily all day. The distance involved made it impossible for the one set of anglers to know how much better the other bank was fishing, whereas even on quite a large lake, the differing success would have been apparent and remedial action easier to take.

It's certainly far more difficult to run a successful lake fishery than people think, especially if, as will most likely be the case, you are unable to drain your lake periodically and start all over again. If you are lucky enough to have a lake which can be drained quite easily, or are in the even more blessed state of being able to build a lake from scratch, then I can only refer you to *The Management of Angling Waters* by Alex Behrendt. He is almost certainly the leading authority in this field, not just from the length of his experience but, more importantly, from the success of his fishery, Two Lakes. His book will give all the information needed on the building of a lake. The stillwater fisheries available to most clubs will almost certainly be gravel pits. Gravel pit lakes are to be found all over Britain, and new ones are constantly being made. If you wish to create a stillwater trout fishery, you will probably turn to a gravel pit, though other types of flooded workings may be available.

FINDING A LAKE

The initial stages of the process are the same as for a river fishery. First find your water, though in the case of some coarse fishing clubs, which already rent several lakes, it may just be a question of deciding which one is the most suitable for a trout water. Our largest fishery started in just the opposite way, however; the big lake was rented with the express intention of starting a trout fishery, and a couple of smaller lakes on the site were taken on as coarse fisheries.

New waters are always appearing, as the demand for building materials seems almost insatiable, but remember that competition for

Fig 44   Trout rearing benefits from a nearby river or stream.

them is very keen. Enquiries made with firms which extract the gravel may help you to find a water, but having the right people to make those enquiries is an enormous asset. If people know each other or do business with each other, they have a head start in negotiations. There are always problems obtaining a suitably long-term agreement, and any opportunity of buying a property or purchasing just the fishing rights and access should be considered very seriously. Grants to help with such a purchase are available from The Sports Council (*see* Useful Addresses). The necessary application forms and guidance leaflets will be supplied on request. How much money may be forthcoming depends on what sort of club is being formed; the more exclusive the club, the less likely the grant. Grants may also be given to help purchase a lease, if it is a sufficiently long-term one. The possibility of buying fishing rights should always be borne in mind, including the rights on lakes or rivers which the club already rents; farming is entering a period of reduced profitability, and some farmers may well be glad of such an injection of capital.

ASSESSING A LAKE

Finding a lake is only the beginning of the problem; now you must assess its suitability.

Water Quality

Water quality is of paramount importance and it is essential to seek the advice of your water authority's fisheries scientist. Some gravel pits are pretty horrible. The water may look beautiful and clear but, for a variety of reasons, the pH may be very low and food scarce. There are three lakes only a mile or so from my house, two are good coarse fisheries but the third, the largest of them and a lovely looking sheet of water, holds no fish and efforts to stock it have been totally abortive; the water is just too acid, and has been in that state for as long as the locals can remember. Unless the scientists give the lake a good report, forget all about it; you will waste plenty of money in any case, without backing a horse which is going to fall at the first hurdle!

Size

Once the water has been given a clean bill of health, its size must then be taken into consideration. Some stillwater fisheries are far too small for my taste, giving the impression of fishing in a goldfish bowl. I like to have the illusion, and remember it is only an illusion, that the fish are more or less wild. Certainly, if the water is several acres in extent there is the feeling that the fish are free and that they can get right away from the angler. A more practical consideration is the number of anglers to be accommodated. A few friends may easily rent a water of about a couple of acres without getting in each other's way and without having the feeling that the water is just too artificial. A club really needs to control a much larger water than this, especially if members are to be free to fish when they like. The system of 'named days' works well enough on some commercial fisheries, but I don't favour the idea on a club water. It is not unreasonable to limit the number of visits which may be paid to one or two per week, if the water is not as big as one might wish. The topic is discussed a little more fully in chapter 5.

Depth

Hand in hand with size comes the question of depth and, even more important, variety in depth. Good shallows encourage weed growth and help to provide a well-stocked larder for the trout but, unless the lake has several areas of deep water, the prospects of running a successful fishery are poor. For a start, high water temperatures are deadly to trout. The fish need cool areas to which they can retire when the weather is hot. It is very helpful to have parts of the lake which are four or five metres (15–18 feet) or more deep, and doubly helpful if such water comes close to the bank as, not only do the fish find a refuge from the heat, but the angler is guaranteed areas of weed-free water. Weeds, beneficial as they are, can be the curse of still waters and, at worst, a lake can be almost unfishable from June onwards. Areas of deep water also help the fish to overwinter well; I was interested to learn that, in one large reservoir which is run as a coarse fishery, echo soundings showed that almost the whole fish population crowded into one area of deep water once the weather had become really cold. There is no difficulty in checking the depth of a possible water; coarse fishing rod, float and plummet will soon give quite a good picture, especially if you take a little boat and row round, though merely casting from the bank will give plenty of information. The more sophisticated use an echo sounder to do this; one of our number, who has been described as a 'dedicated fanatic', actually owns such a device and has built up a good picture of the lake for us. We fear an ulterior motive; he just wants to locate possible hot spots for one of our monster pike!

Extra Water Supply

One other desirable feature, though this is by no means essential, is a fair sized stream, or even a river, close at hand. If you have any plans to rear your own trout, at some later stage in the project an extra supply of water becomes very important indeed. Fish rearing is discussed more fully in chapter 6. Just in the normal course of events, however, it is very beneficial to the fishery to be able to pipe a large volume of fresh water into the lake and draw water off at the other end. This has been done at two of our lake fisheries and a significant improvement in the fishing has been recorded, especially later in the season. Provided that lake and stream are not too far apart and that there is a fall in level from stream to lake, this is not a difficult undertaking. The inflow is attractive to the fish, as food is carried along by the current, and it should be a source of fresh, well-oxygenated water. It certainly seems beneficial to have a change over of water in the lake, always assuming that the water coming in is pure. The disadvantages of a neighbouring river are equally obvious. Every few years the change of water in the lake is much more sudden and dramatic, when there is a serious flood. On several occasions in the last twenty years all our lakes, the stream and the river have become one unified sheet of water, just lapping the front step of the fishing hut. Unfortunately, I have no photographic evidence of a really big flood, but even a modest one like that in Fig 45 causes problems when trying to feed the trout. We had a huge flood one year just after our main stocking, in fact it was on April Fool's Day, which seems singularly appropriate! We lost nearly all our stock of fish, especially the rainbows, though people further downstream had no complaints! Despite this sort of inevitable disaster, I am certain that the overall gains far exceed the losses, so if a good supply of running water is available, that is a strong plus in favour of the site.

Fig 45   Sometimes there is too much extra water. This is the same rearing pen seen in Fig 44, during a big summer flood.

67

Access and Services

I have mentioned a few of the features to look for to suit the fish, but the fishermen also need some consideration. It's not much use being able to rent a beautiful lake if you can't get to it – convenient access is most important. In general, there will be no great problem with gravel pits, as lorries and machinery needed access when the pits were being worked. Fishermen in general are usually prepared to walk, but it is much more convenient and secure to be able to leave your vehicle right on site. Good access is also important to allow delivery of trout and, if you rear your own fish, trout food.

At the same time as you consider access, you should also think about what other services may be available. I'm not saying that gas and mains drainage are needed, but electricity can bring enormous advantages, especially if you are running a fairly large and progressive club. On one fishery, the electricity board made a supply point available and the club laid a considerable length of cable to bring the power on site. The job was well worth the labour and expense. It's not just a question of sitting by the electric fire, brewing up tea and cooking a hot pie, or even enjoying the mod cons of a flush toilet; the power makes the actual work on the lake much easier. Pumps and aerators have been installed at this particular fishery to help with the fish rearing, and we even have a fridge to keep the fish fresh when we have caught them!

The Banks

The state of the banks is probably the last of the features of a lake to be taken into account. On one of the fisheries we rent, the owners had left the banks beautifully graded and grassed, so that it was a pleasure to fish from them, but other waters may be in a sorry state. Some have become overgrown with trees and brambles, others may have high banks, which will be very uncomfortable to fish from. If the other features suggest that the lake may make a good fishery, these are quite simple obstacles to surmount, provided that the support of the owner is forthcoming. It is vital to stress this point, because the owner's interests may clash with what is needed to run a good fishery. If he wants a wilderness where his pheasants can hide, then the prospects for trout fishing are poor, even though the sporting seasons don't coincide. Any agreement with the owner of such a lake must specifically include a realistic programme of bank clearance and improvement. Where the banks are too steep or too

Fig 46   *My wife fishing the inflow. Note the bank, which is*
protected by bags of sand and cement, and the grid close
enough for easy cleaning.

uneven, the cost and feasibility of bringing in machinery must be carefully considered, as this can add enormously to the expense of the venture. I am not suggesting that everything should be smooth as a bowling green, but a fishery needs regular maintenance and if grass cutting, for example, has to be done by machinery, reasonably smooth going makes work much easier and cheaper.

This should cover the basic factors by which to decide whether or not to rent a lake. Just drop in a few trout and I suppose you have fishing – of a sort. To make a real success of the venture, however, sights need to be set very much higher. Unless you have been very lucky in your choice of water, a continuous programme of maintenance and improvement will be needed. Work can be done to improve the quality of the water, more so in a lake than in a river, and this should improve the quality of the food supply at the same time. The banks must be kept in good repair, as erosion may well be a problem if the lake is a fairly large one. In conjunction with this work, extra fishing spots may be provided, by giving anglers better access to the water. The quality of the stock fish must always be considered and the advantages of rearing your own trout are discussed in chapter 6.

Finally, there are the amenities for the members to consider. These are far more important on a lake fishery than on a river, and increase in importance directly in proportion to the size of the fishery. I will discuss amenities more fully in chapter 5. There will be no shortage of work and, I hope, of helpers, but it is wise not to attempt too much too quickly. The number of helpers will diminish rapidly if they find that they are expected to spend more time working than fishing.

WATER QUALITY AND THE USE OF CHALK

I have used chalk in varying forms both in our river and in one of the lakes. I certainly think that it is beneficial, but to what extent is hard to quantify. When working on the river, for example, we have put in ordinary chalk as delivered to the farm; Nautex has been used, but it is so expensive that you would think you were putting gold dust into the water. Luckily, Siltex has been introduced, which claims to be almost identical to the French product in everything but price. There is absolutely no doubt that vast quantities of mud have disappeared from my length of the river and there are now long stretches of gravel which certainly haven't been exposed within living memory. The problem in analysing its beneficial effects is that the chalk has been used in conjunction with all the other improvements, such as the use of groynes and the narrowing of the river; both of these measures speed the current, so that it carries silt away more readily. It is therefore not possible to assess the full value of the use of chalk. One of the French papers I have been reading states that there was a 13 per cent reduction in the depth of mud on the treated stretch – but that there was an almost equal build-up further downstream.

To date, we have used only ordinary chalk in our lake and not one of the chalks marketed specifically for fisheries work. The whole operation was a pretty haphazard affair; a great load of chalk was dumped on the bank and then shovelled into a couple of boats, which went to and fro spreading the stuff. The high spot of all this activity was the slow sinking of a leaky and grossly overloaded punt, and the consequent immersion of its occupants. It was a cold winter's day and the swimmers gave no report on the fish stocks which they had observed while under water; they seemed much more concerned with getting back to the hut and changing their clothes! Anyway, I think that the chalk may have done some good, though I wouldn't swear to it. We have not repeated the treatment as the water quality in the lake is very good and we worry that more chalk may stimulate the already luxuriant weed growth. Chalk has also been used, in a much more solid form, to make a firm bottom in a couple of the rearing pens; the water in them is always beautifully clear over the chalky bottom and the chalk should help in the breakdown of organic matter. The use of Siltex in one of the large fish-rearing bays is being considered with the express intention of hastening the breakdown of organic matter. I believe that the use of chalk *is* beneficial in still water, but I am a bit non-committal, partly because my experience is limited and partly because our lakes already have excellent water quality.

Having embarked upon the subject of chalk, it would be wrong of me not to discuss a little more fully its use and the advantages claimed for it. The lake Frank Sawyer was one of the first keepers to make extensive use of chalk to improve his fishery. Like most prophets, he was largely ignored in his own country and it was left to the French to look into the matter more scientifically and to develop a product specifically for fisheries work. Sawyer's observations suggested that the addition of chalk caused a considerable reduction in the amount of mud in his stretch of the Wiltshire Avon. He also claimed that chalk made the food supply much more palatable to the fish and this, in turn, led to an increase in the average weight of the wild trout. Calcium helps various crustaceans, crayfish, shrimps and the like to flourish, thus increasing the available food supply, a supply which produces pink-fleshed trout. Exactly how the chalk works is a bit of a mystery to a non-scientist like me. The French papers talk about 'colloidal flocculation', which could mean anything, as far as I'm concerned! I gather, at its very simplest, that the fine chalk particles penetrate the top layer of silt, allow oxygen to enter and then the bacteria can set to work to break down the organic matter. If the silt is deep, only the upper layers are affected, so top-up doses of chalk are needed until the good work has been completed. Even in a lake, therefore, there will be a reduction in the amount of organic matter covering the bed. The conclusion reached in the more objective of the scientific papers is that there is a limited but definite reduction in the depth of the silt, and that there is a positive improvement in water quality.

Apart from the beneficial effects already mentioned, an application of chalk should increase the pH of the water making it more alkaline. There is nothing new in this. For very many years acid lochs in Scotland have been fertilised and limed in order to improve the fishing. Most recently, a scientific project has been carried out at Loch Fleet in Galloway, so that results could be properly monitored. In this case, the land surrounding the loch was treated with lime, so that the run-off after rain would be much less acid. The pH rose from pH 4 (very acid) to pH 6 and trout can now live in the loch. Similar work is being done to counteract the effects of acid rain.

In France, chalk has been used to improve the water quality in lakes which have been polluted. Considerable success has been achieved, but I do not believe that it is quite the panacea that some claim it to be. In many of the lakes, other work has been done to reduce the amount of pollution entering the water, so the chalk cannot claim the full credit. Even so, there is definitely improved pH and better-dissolved oxygen content. The growth of rooted weeds is healthier and the water much clearer. To the best of my knowledge, chalk has not been much used in this country to help the recovery of polluted lakes. By the time this book is published, however, a considerable amount should have been used on an experimental basis in some of the Norfolk Broads, where the coarse fishing is in a sorry state of decline and water quality leaves much to be desired.

Before using chalk, it is wise to seek the advice of your fisheries scientist and also the advice of your supplier. As a general rule, about a tonne to the acre is used for the main application and the more widely it can be diffused in the water, the better. The chalk marketed for this kind of fishery work is extremely fine and is made up of microscopic fossils called coccoliths (*see* Useful Addresses for suppliers).

As a postscript to all this scientific talk, how illogical the question of water quality can be! Limited pollution can be quite beneficial. As youngsters, my friend and I thought it a special treat to get on the train to Bradford-on-Avon and fish in the 'sewer hole' there. We used to catch some splendid roach!

CREATING AN EXTRA WATER SUPPLY

Another way of improving the water quality in a lake, or at least modifying it, is to provide an inflow from a nearby river or stream and to draw the water off at the other end of the lake. This allows a gradual change of water and I am convinced that it is very beneficial to the fishing. Though the job is basically simple, it can be fairly expensive; machinery is needed to dig the trench into which the pipe will be laid, and the pipe itself, about 30cm (12in) in diameter if possible, is not cheap. The actual job of laying the pipe is for your contractor to do and

Fig 47   The inflow on a lake. It extends too far from the
bank for easy maintenance and there is no grid to deter
escapees.

should present no problems, though on one of our lakes an absent-minded digger operator managed to cut a channel from river to lake and then realised he had left his machine with one set of tracks on each side of the trench!

The water levels must first be checked, before any work is attempted, in normal summer conditions because that is when the fresh water is most useful; fortunately, river levels are often a bit higher than normal in summer as the weed growth holds the water up. You need a reasonable drop, 20cm (8in) or more, from river to lake at the inflow and, obviously, another drop at the outflow, from the lake back to the river, or to any other suitable adjacent water course. If the levels are right, then permission to do the work must be obtained from the water authority. This should be a formality provided that the water that has been abstracted is returned fairly quickly to the river. Finally, before embarking on the work, do go into the cost very carefully.

Once the pipe has been laid and the soil replaced, several other jobs must be done to complete the project successfully. It is important to make the inflow pipe in the river secure from erosion (*see* Figs 48 and 49). The easiest way to do this is probably to reinforce the bank with a wall made of bags of sand and cement. This prevents the pipe from sucking in mud and eroding the river bed, and also provides firm

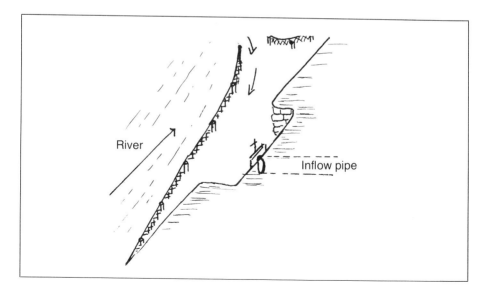

Fig 48   Lake inflow pipe leaving river.

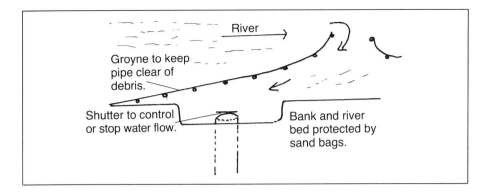

Fig 49 Inflow pipe in river, seen from above.

standing, if any more work needs to be done at the inflow. On one fishery of ours, the inflow is from a fair-sized river and the wall has been further protected by some metal sheets which help to deflect the current and to keep debris away, so that there is less risk of blocking the pipe.

The final job at the inflow is to provide a shutter, so that the amount of water entering the lake can be controlled, or the pipe can be completely closed. This is necessary in order to prevent silt-laden water coming into the lake when the river is in a dirty flood or, worse still, to keep out pollution. A flat metal sheet to cover the river end of the pipe is all that is needed. A technically-minded member will easily devise a way of raising and lowering it.

There is much less risk of erosion where the pipe enters the lake, but a few bags of sand and cement will help to give a neat finish to the job. More important is the provision of a good grid, so that your trout can't emigrate, as they most surely will if given half a chance. This is another technical job, so delegate it to your shutter specialist! Even though the inflow is protected from the worst of the debris floating down the river, a certain amount of weed and rubbish will find its way down the pipe, so the grid over its end should be slightly inclined, in order that debris will be pushed to the surface. The bailiff will need to clear the grid every day, though we always leave a rake handy, so that any club member can do it *en passant*. The grids on the outlet pipe from lake to river should also be cleared daily. Coarse fish fry may well enter via the pipe, but the lake is almost certain to hold a fair coarse fish population in any case. Pike and competing fish are mentioned in chapter 7, but one of our lakes has been a trout fishery for over twenty years, and the coarse fish are really no problem.

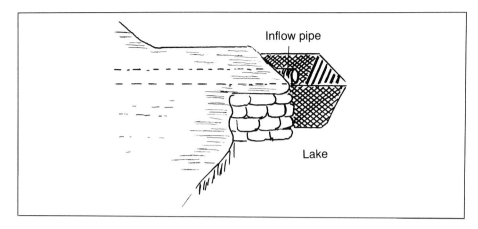

Fig 50   Inflow pipe entering lake.

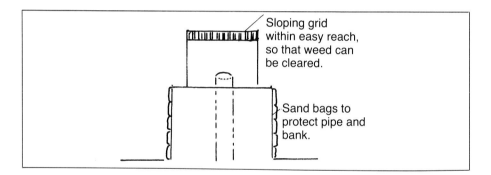

Fig 51   Plan of inflow entering lake.

WEED CONTROL

Before leaving the subject of water quality, weed growth and control in the lake need to be considered. Most trout waters have quite luxuriant growth and this helps to create a healthy environment for both the fish and their food supply. Some years, however, things get totally out of hand and the angler finds that he has nowhere to fish. Cutting the weed is a possible solution, but not a very satisfactory one. To start off with, it is very hard work, something to be avoided if at all possible, and the cut weed must be dragged ashore, which is even harder work, otherwise rotting vegetation will de-oxygenate the water. In any case, the weed will grow again at an alarming rate, so the work will have been of only limited benefit.

Alex Behrendt describes the use of black polythene sheeting to shut out the light and thus prevent growth. A friend of mine, who looks after a neighbouring fishery, gave it a try, but found the polythene rather unwieldy to handle, so that the results he obtained were not particularly successful. On our biggest lake, we are lucky enough to have a couple of areas within casting range which are deep enough to remain weed-free. This means that there will always be some fishable water but, when it seems that a weed explosion is likely, we treat a few shallower parts with herbicide. By shallower I mean fish holding water up to about 2 metres (7 feet) deep where, if not treated, the weed would reach the surface. If possible, I would avoid the use of herbicides as, despite all the claims made for their safety, I cannot believe that 'poisons' can be good for the water. Even so, management means taking decisions, and if the weed growth looks threatening it is best to treat selected areas early, to ensure adequate fishing space. The manufacturer's instructions must be strictly followed and the weed should be treated just as it is starting to grow, so that there will be as little rotting weed as possible in the water. Finally, permission and advice should be sought from the water authority. I keep mentioning this, but it is pointless to have experienced scientists available and not to make use of their knowledge. The herbicide we have used is called Midstream, and carefully-selected areas have been treated very successfully. The expense is quite sufficient to ensure that no more water is treated than is necessary!

To underline the need for caution when using herbicides, one local fishery in Norfolk was quite badly polluted following the incorrect use of chemicals. The instructions had not been followed and far too great an area was treated. The weed problem was eradicated, as were many of the fish, because the mass of rotting weed de-oxygenated the water. It was necessary to leave the lake to recover and a year or two later the water authority came to the rescue with a weed-planting operation. Luckily, the damage does not seem to have been permanent and the lake is running again as a moderately successful fishery.

Cutting weed and dragging it out is hard work, and not very successful either; black polythene sheeting is awkward to handle; herbicides, which can do a very good job, have an element of risk in their use. So one wonders if there is any other remedy on waters which have a serious weed problem every year. Observation has clearly shown that areas of deep water are always reasonably weed-free so, on waters where such areas don't exist, it should be possible to make some, provided the owner of the fishery is agreeable. Such drastic measures would make even the regular use of herbicides seem inexpensive but, in a well-established club, the expense might be worth considering. Apart from providing permanent weed-free areas within casting range, the deeps should also provide cool water to which the trout can retire during one of Britain's rare heatwaves. One point – this is the only management idea described in this book which I have not yet tried.

BANK MAINTENANCE AND IMPROVEMENT

Control of aquatic weed leads, fairly naturally, to control of the banks and what grows on them. You must make and maintain a good, firm track all round the lake, not only to allow access for the angler, but also to enable materials to be transported to wherever work is being done. Once the track has been made, it must also be kept open. Grass cutting is discussed elsewhere, but I'm a great advocate of sheep for this job.

Materials

We have found that the materials most useful for bank maintenance on our big lake are bags of sand and cement. I am happy to make do with plastic sacks filled with earth for much of the river work, because there is no access, except on foot, to many of the places where I am working, and even a weakling like me can stagger along with a spade and some empty sacks! However, it is best, especially if there is a lot of work to be done, to have a central base with sand, cement and cement mixer available and to fill the sand bags there. Mix the sand and cement *dry* and do not over-fill the bags; they can then be pummelled into the shape required, and you are much less likely to do yourself an injury when putting them in place. Sand bags are not cheap but, if a good job is done, it is money well spent. The finished products can easily be transported to where they are needed, either round the perimeter or by boat. Once they have been laid in the water or rained on, the bags will quickly harden and adhere quite firmly to each other.

Protecting and Improving Promontories

The need for bank protection varies enormously, depending on the size of the water and the nature of the banks. An exposed lake of fifteen acres or more can get very rough in a storm and considerable erosion may take

Fig 52 Promontory built in shallow water, using bags of sand and cement and then filling in with rubble and top soil.

place, making access to the fishable water progressively more difficult. Even on a much smaller water, work will almost certainly be needed to protect promontories and islands from which the members fish. Protecting a promontory is simple enough; just build a wall of sand and cement bags all round the most exposed part.

A couple of improvements could also be made. Firstly, if the water round the point is shallow, it is worth extending the promontory so that more fishable water is made available. Build the wall of bags as far out into the lake as desired, filling in the shallow water with earth, rubble, or whatever materials may be available. The second improvement,

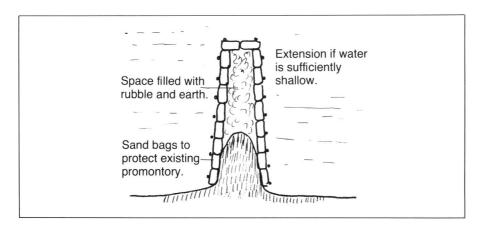

Fig 53 Promontory strengthened and extended using sand bags, seen from above.

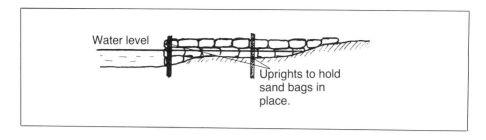

Fig 54   Side elevation of promontory extended using sand bags.

especially useful if the promontory is being extended, is to drive in stakes, so that the bags won't fall or be pushed out into the water.

Another, but less attractive, way to protect this sort of fishing spot is to make use of old tyres, which are usually freely available. They are not very easy to use, as they are quite likely to go floating off down the lake before they are securely in place. A few stones will soon cure this tendency and putting the tyres over a small stake, hoop-la fashion, will help to prevent subsequent defection. The wall will only need to be two or three tyres high; fill each pile with rubble as you go along. When this part of the work is finished, a telegraph pole, or suitable tree trunk, is laid on top and held very firmly in place, by stakes driven into the bed of the lake. The whole area should then be built up with rubble and earth to the level of the poles or tree trunks. There may be some erosion caused by waves breaking over the point, if it is in a very exposed position, and an extra load of earth or rubble may be needed to make good the losses at some future date.

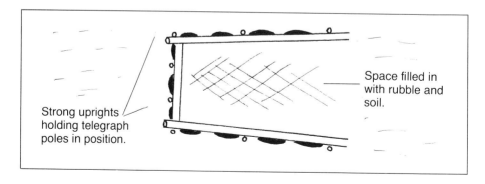

Fig 55   Promontory strengthened and extended using tyres and telegraph poles, seen from above.

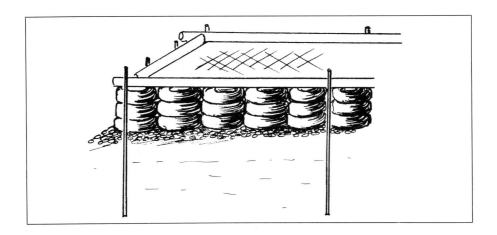

Fig 56 Side elevation of promontory extended using tyres.

In general, this kind of job is very secure from erosion and the point can easily be extended using the same methods if the water is not too deep. Grass seed, or some turfs if they are available, will put a good finish to the job, though this is best done at the end of the season, so that it will have a chance to become established before next season's onslaught. Although this is not the prettiest of jobs until it has had a year or two to naturalise, it is very secure and effective. Good access is vital before attempting work like this, because a large amount of earth and rubble needs to be dumped close at hand.

Fig 57 A very exposed point, protected and extended using tyres and telegraph poles.

81

Fig 58   Bank protection
work using old wooden
and galvanised telegraph
poles.

Fig 59   Bank protection
showing how rushes grow
up round telegraph poles;
note the large fish trap
which is worked from the
bridge.

Rushes

Where a whole shoreline is subject to erosion, other measures are more useful and cost effective. A good thick bed of rushes provides excellent protection, but the problem is to get the rushes to grow strongly enough before erosion can wash them out. Our solution has been to stake telegraph poles or tree trunks along part of the threatened bank and to plant rushes between them and the bank. The poles protect the rushes from the initial battering by the waves, and give them time to grow and spread. After two or three years, the rushes should be thick and strong and the poles may be moved to carry on the good work elsewhere. Suitable large and straight poles may well be hard to come by, but in the south of Britain, the great storm of 1987 should have ensured that nobody is short of timber for years to come!

Once thick rush beds have developed, of course, more work needs to be done, because the angler can't really get to the water to net his fish. Fishing platforms need to be built at intervals so, more bags of sand and cement! This is a simple enough job; the major problem is what to put on top of them for the angler to stand on. We got a load of iron gratings,

Fig 60 Fishing platform in
rush bed, made from iron
grating.

Fig 61 Walkway through
a reed bed, using old wire
mesh frames.

about two metres (7 feet) long and a metre (3 feet) wide. It took a bit of
man-handling to put them in place, but they should now last for ever,
and they are non-slip into the bargain! If you have to have wooden
platforms, make sure that the wood is well treated and good quality,
otherwise the whole lot will rot in five minutes. Don't forget to fasten
wire netting to the top, so that the angler is less likely to join the fish in
the water. Hold the platforms firmly in place by fastening them to poles
hammered into the lake bed, and then they won't go floating off in the
first flood.

Islands

Offshore islands, provided you have sovereignty over them, should be
brought into use as much as possible, though they may need some of the
treatment just mentioned in order to prevent their disappearance; in the
course of twenty years it is amazing how many little islands have
vanished from our biggest lake, leaving a dead tree or two sticking out of
the water to mark their last resting place. Some islands may need no
more than a telegraph pole bridge, or some equally simple structure, in
order to reach them, but the more exotic shores must be reached by

boat. We actually built a couple of islands on distant shallows, using four oil drums filled with concrete to form the base. Though these islands have been a great success, I think it is much simpler to provide a few buoys to which boats could tie up, thus enabling water which would otherwise be inaccessible to be covered. I would certainly not permit fishing from the drifting boat, except on very large waters; even on the reservoirs, some idiots in boats can still get in the way of bank fishermen! If boats are to be used, it is essential that they are stable; we have one disgusting little craft which is always threatening to tip its occupants into the water – and it has actually succeeded on some occasions, so make sure that the club insurance covers all eventualities. Fibreglass seems the best material for the boats as it needs only the minimum of maintenance.

All the jobs just mentioned not only help to keep the banks in good order, but they also provide that degree of variety for the fishermen which I have advocated elsewhere for the fish. To be able to offer a wide choice is a great asset for a fishery – a sheltered bay, an exposed shore, a promontory which commands a lot of water, a platform hidden in the rush beds – if all are available members should always find something to suit their taste. The practical advantage is that, with a large number of different fishing places available, there should be less risk of over-crowding, especially in awkward fishing conditions. I'm quite a sociable chap and am quite happy to fish a lake with a friend on either side, but a long line of fish-frightening water floggers, looking as though they were pegged out for a match, is definitely not to my liking nor, I feel, to the liking of the majority of our members.

TREES

When considering the banks of the fishery, the management of trees is also important. I think the 'chop 'em all down' brigade, and that includes some water authorities, has at last been discredited and that more sensible policies now prevail. Unfortunately, one also has to contend with the other sort of fanatic, who thinks it a crime even to touch a tree, let alone cut it down.

As I mentioned earlier, willows need regular pollarding, if they are to live long and useful lives. Alders should be prevented from growing too high, and elders benefit from regular pruning. These trees are almost synonymous with the waterside, and they enjoy the damp conditions greatly. I'm told that the alder was so prolific at one time that it was on the list of notifiable weeds. Observation will show what other trees may grow well in the vicinity of your own water; the old-fashioned English country house lake, surrounded by well-drained parkland, may be suited to some of our most beautiful deciduous trees, but these trees would have miserable prospects in the damp, peaty soils which surround many old gravel pits. In any case, I prefer a stillwater fishery which is not too sheltered from the wind; the trout are often quite hard enough to catch, without having to contend with a glassy calm every time you go fishing!

I am not persuaded by the argument that trees are as important on a lake as they are on a river to the well-being of the trout. They offer a little shade, it is true, and the odd titbit falls from the branches, but trout in still water rove widely to find their food – they can't loaf about like river fish, waiting for the current to bring them their dinner. Nor are trees very important in helping to stabilise the banks; in fact, planted too close, they soon fall in and take a great chunk of bank with them. Those that don't fall in manage to drop a generous helping of unwanted branches into the lake, and their removal means extra work for you. Not that I would have you believe that I am anti-tree; quite the reverse, but round the lake I believe the trees are there for their amenity value, for the benefit of the fishery rather than the fish. If you agree with this view, then trees should be planted with the *angler* in mind or, more exactly, the angler's casting. On one fishery, we have a maddening stretch of bank, with another lake barely twenty metres (22 yards) behind it. The moment you have a lapse in concentration when casting, you are 'caught behind' by the trees surrounding that lake. I wouldn't mind a fiver for each of the flies I've left in the trees surrounding the other lake! Trees which are not to be planted at the lakeside should be set well back so that they don't constitute a hazard. It is only too easy to forget just how big a tree grows; that little sapling is soon quite a large tree, with an insatiable appetite for flies.

Trees growing at the water's edge always seem to have extra beauty, but they also have a practical use in breaking up lengths of bank, thus giving individual anglers more privacy. If the lake's shoreline is very irregular this is not a very important consideration, but where the banks are rather straight the trees will make an enormous difference. How trees may be planted depends on the owner's wishes and the amount of land available. A little clump of trees and bushes makes a much more effective screen than an individual tree and, if one tree falls down, all the good work is not lost. I certainly prefer planting like this, if there is sufficient room, and then leaving 20 metres (22 yards) or more of clear bank so that the angler has plenty of casting space and the breeze can get at the water.

I would prefer to draw a veil over our own tree-planting efforts, as we have not had spectacular success. In fairness, the land round our biggest fishery is rather poor, just wet, peaty stuff; even the sheep don't think much of the grass and, higher up, we seem to have almost pure sand, giving a spectacular show of broom and gorse but little else. As far as we are concerned, the most successful trees are, inevitably, willow and alder. Several areas well back from the lake were fenced off from the sheep, in an effort to provide more variety in the form of little plantations. Silver birch seemed to do quite well, but they were only planted to provide a bit of shelter for the other trees. A few conifers struggled upwards and, elsewhere, a line of poplars made an effort, but forestry experts have not come flocking to discover the secret of their astonishing growth rates. Hawthorns seem to do quite well; they are quite useful at the waterside, as they don't grow too large but still make a good screen, and their flowers and berries are very pretty to look at. Some years, there is a sufficient hatch of hawthorn flies to attract the attention of the trout, should enough flies fall on the water. Hawthorns also harbour an unfailing, almost magnetic, attraction for the artificial fly; it's marvellous how your leader can play cat's-cradle round the very thorniest bits. I hope that you will be blessed with better soil than we are; your own observations will tell you, much better than I can, what trees will do well in your own area – happy planting!

The results of your labours attempting to improve stillwater fisheries may be somewhat less spectacular than the transformation which can be brought about on a neglected river fishery, but steady work is important to maintain a good standard of fishing; many lakes have started well and then declined year by year. Sensible work should avoid this decline and, as your skill increases, lead to an improving fishery. Set-backs will almost inevitably occur, but it is not realistic to expect perpetual immunity from eye fluke or some other pestilence and our delightful climate may deal some nasty blows. Even so, if you have a loyal club membership and a good team of helpers, success should come your way most seasons – and I hope it does.

5
Looking After the Members

Fishing varies enormously between that found in lowland regions and that of the uplands. Lowland fishing tends to be conducted in a domesticated, closely-managed setting, otherwise there would not *be* any such trout fishing. Conversely, a day's loch fishing may well entail a couple of hours' walk each way before the chosen water is reached; there will probably be a stop or two *en route* to have a rest and to consult the map, or even to have a cast in other lochs. The hike itself is part of the adventure. Your only company may well be a few sheep and a distant view of some deer, with no sign of roads and civilisation. The upland rivers are equally wild, one moment in spate, the next back to normal. In some remote areas, if you walk far enough, there is every chance that the water may not have been fished at all that year.

In the lowlands, the angler may scarcely need to walk at all, finding a car park almost at the water's edge. The waters are surrounded by sheep and cattle, crops and tractors – by civilisation, in fact. The lakes may well be man-made reservoirs or gravel pits, and the rivers have often been controlled by the building of weirs and mills. In my view, the fishing provided in the lowlands should mirror the familiarity and domestication which is to be seen all around, and the amenities provided for the fisherman, wherever the water may be, should be appropriate to the setting.

CAR PARKING

On arriving at a fishery the fisherman has to put his car somewhere. Access and parking must be agreed with the owners, but there is seldom any problem on a gravel pit fishery, as access is normally across good firm ground. If it is not, however, the site must be prepared, even if it means spending hard-earned money on a few loads of hard-core – you don't want to waste good fishing time getting Albert's Mini out of the mud! Life by a river is much less frenzied than on lakes – the river man is more of a loner – but even here, the anglers must have a clear agreement with the owners as to where cars may be left; constant obstruction of farm work will lead to the rapid loss of a fishery.

LAKESIDE CLEARANCE

Recently, I visited a couple of gravel pits managed by a friend of mine, and the state of the banks was a revelation. It would be an exaggeration to say that beautifully-manicured lawns stretched right down to the water's edge, but this was very nearly the case. My friend is a generous man, and he very kindly allows a firm who manufacture mowing machinery to test their products on his grass. I doubt whether you will be lucky enough to find such a firm so conveniently situated, but it is important to provide good access to the water for the fisherman. Where the banks are well graded, it is easy to bring in a tractor and keep the grass cut by machinery. This is what we do on one of our lake fisheries. How often the cutting is done depends on how much you are prepared to pay and the standard of appearance you wish to maintain.

On another fishery, we use the 'Ovine Nibbler'. The 'Nibbler' works seven days a week, doesn't break down or go on strike, and has very low running costs. Certainly, if conditions are suitable and the owner allows it, I think that a small flock of sheep is a splendid idea. In our part of Britain, people are always on the look-out for a bit of extra grazing. I prefer sheep to cows, as they do much less damage to the banks. Whatever method you adopt, the banks round a lake fishery need to be maintained in an attractive way.

RIVERSIDE CLEARANCE

Access to a river can entail much more work, depending very much on the type of agriculture practised. If cattle and sheep predominate and are allowed to graze right to the water's edge, then all the clearing away is done free of charge – and effort! If the water has not been used much and there is plenty of woodland around the river, then it is a very

Fig 62   'Ovine Nibblers' taking time off from their work.

different matter and, in the first instance, it may well be a question of jungle warfare. Before embarking on any clearance programme, however, it is essential to have the landowner's permission, especially when it may be necessary to fell some trees. If you are lucky, you will not only have the owner's permission, but also his whole-hearted support; one of our landowners most generously made a strimmer available, an expensive item of equipment, but very useful indeed to maintain clear paths. On a serious note, you only have one pair of eyes so don't strim without wearing the goggles provided. It may be very irritating on a hot day, when they steam up and sweat runs down them, but twigs and stones can fly up at frightening speed and it is folly to take risks.

When there is major clearing to be done, it is a question of striking a balance between the anglers' requirements and those of the fish. As I have already suggested, overhead cover can do much to enhance the fish holding capability of a stretch of water, so it is essential that the fish are left with an adequate amount of shelter whilst, at the same time, allowing the fisherman ample room to walk and fish. The path itself should certainly be cut at least a couple of metres back from the water, so that the quiet fisherman can pass without frightening the trout. It is just as well to make quite a wide path while you are about it, as it is surprising how quickly new vegetation will encroach. If the path has

Fig 63   The path by the water, with a screen of vegetation left to hide the angler.

become badly overgrown, saw, slasher and pruner will all be needed in the early stages of the work, but a chain saw should only be used by a competent operator. A thorough clearance job done at the outset will certainly make maintenance easier later on. When both banks are wooded, the problems multiply; not only will more clearance be needed but the river may well be too dark and shaded to be really productive. One bank will need to be quite ruthlessly cleared, again subject to the owner's permission.

On fairly small and shallow rivers, it is quite likely that there will be a barbed wire fence to stop the cattle crossing from one bank to the other, especially when both banks are not in the same ownership. This stretch of no-man's land is a mixed blessing. The banks of the river don't get trampled and turned into swamps under the animals' feet, but the vegetation may well run wild. This means that it will need strimming several times during the season in order to maintain a good footpath. However, a little hard work at the right time can improve the situation greatly. I find that most farmers are very helpful, especially if you are prepared to lend a hand to help them also. When work on their fences needs doing, take the chance of positioning them to your maximum advantage, which I reckon is two to three metres back from the water. This means that the cows can graze almost the whole of the strip for you; it's surprising how far they can reach through the wire and still leave a good fringe of bankside vegetation to conceal the angler. On stretches where the fence is satisfactorily placed, strimming can be reduced to an absolute minimum, in fact I get away with just one quick cut during the season.

Time and labour will not allow all the work to be done at once. The first priority is merely to provide a reasonable path. The need for some jobs may only become apparent after a year or two on the water. Other improvements to the path may not be essential, only desirable. Not all your members will be young and active; I fear that age and arthritis take their toll on all of us, so look on some of the jobs as your insurance policy! There was one section of path on our water which was not only a bit of a hill, but also sloped unpleasantly toward the river. I set about improving it one winter; only about fifty metres (55 yards) needed attention, possibly a little more. I cut down a few straight alder trunks to form the riverside edge for the path in the worst places and held them in place with a post at each end. I dug the path level, jamming the soil which had been dug out against the alder trunks. I also cut a few steps to make going up and down the hill more comfortable. The whole job did not take many hours and the improvement was much appreciated. Low-lying and swampy spots can also be built up to make easier walking. There are several places where I have pulled mud out of the river with the chrome in order to raise the bank. As the mud dried out and grassed over there was a significant improvement. Some spots are always wet and boggy but a few boards, staked firmly in place and covered with wire netting, can make the going much more pleasant.

Having made a good path, maintain it. If you have quite a lot of grass-cutting to do, as a general rule it pays to cut it about three times during the season, so that the grass never gets too long. Stinging nettles and most other weeds cut easily with the strimmer, but long grass is hard work and has a talent for wrapping itself all round the machine. If the banks are sufficiently firm and level, they can be cut more easily with a rotary mower, if one is available. On one of our fisheries, the gamekeeper looks after the banks. One wide section he cuts using the tractor, but he has a 7 hp Hayter to do the rest. He was ill a year or two ago, so I did the job for him. On the really level parts I almost had to run to keep up with the machine! It is a really headstrong beast and I know that on various occasions it has charged off into the river. Despite these minor failings, it makes light and quick work of the grass cutting and beats strimming all hands down.

BARBED WIRE

In the course of this chapter I have used two rude words – barbed wire. I stated my case at the outset; we are dealing with managed, lowland waters, and the angler should not be subjected to an obstacle course. Where barbed wire has to be stepped over, it should be disarmed. Waders cost a small fortune nowadays, so why risk puncturing them? The fisherman does not wish to leave parts of his clothing hanging from the wire. The job of taming the wire is easily done by wrapping one of the ubiquitous plastic sacks firmly round the wire and tying it in place with the equally common red baler twine. A slightly neater and possibly more durable job can be done by slitting hosepipe lengthwise and putting it over the barbed wire, though this is a more fiddly and time-consuming operation. In sections where the river bank has been wired off to prevent the cattle getting into the river, it is essential to provide plenty of places where the angler can step in and out safely.

GATES AND STILES

Apart from places like this, the fisherman also needs to move from one field to the next. Gates can be a curse, as someone is always leaving them open and, rightly or wrongly, the fisherman usually gets the blame. Nor are farm gates ideal passing places in fields where there are plenty of cows. I still remember one particularly bad gateway in Somerset, where I am sure some club members must have sunk without trace in the mud and ended up on the list of missing persons. Some gates are not only padlocked but wired up as well, as a precaution against cattle rustling – not just a figment of the imagination from some television Western, but a reality in many areas. Some gates are so high that climbing over them is a major feat of mountaineering so if possible a good strong stile, set well to one side out of the way of the mud, is a better bet. One such stile is shown in Fig 64, a splendid job all in oak! This one bypasses a gateway which is particularly bad in wet weather and, after severe floods in 1987, my wife and I mounted a major fish rescue operation in that very spot. We scooped out loads of sticklebacks and a few loaches from the large puddle left once the floods had receded, and restored them to their home in the neighbouring drainage ditch.

Fig 64   An excellent stile, well to the side of a very muddy farm gate.

Fig 65   A bridge with rather worn wire netting – time I did some work here.

The mention of stiles brings me to one safety item which will be relevant in the next section also. All horizontal pieces of wood should have wire netting stapled firmly to them, in order to avoid the risk of horizontal anglers. Once wood has become wet and muddy and bits of moss and algae have started to grow on it, it becomes extremely slippery and dangerous. Stiles, bridges, even a plank over a ditch, should all be made safe. It is a simple precaution and costs very little in time or money. I frequently go beating on one estate totally lacking in such refinements, and crossing the ditches on a wet day is risky business.

BRIDGES

Ditches have to be crossed, and we are not all long-jump champions. I can still remember doing a no-jump and starting off in the ditch! Railway sleepers are beginning to get scarce, but they are especially useful for bridging ditches. Wooden pallets also do the job quite well, provided that a couple of good strong lengths of tree trunk are dug into the ground to support them. Always use thick, strong timber for these jobs so that they won't need doing again in your lifetime.

Many people fail to appreciate the fact that rivers do actually flood from time to time. All bridges and planks should be secured firmly in

*Fig 66   Railway sleeper over a small ditch with a handrail –
great for leaning on.*

place, so that they won't get washed away. To hold a plank in place I hammer four bits of wood into the ground, one at each corner, and drive a nail through into the plank. That's quite sufficient to stop it floating away, though a bridge would obviously call for sterner measures. Most of our bridges, even the little ones over the ditches, have handrails in addition to the wire netting, to make them as safe as possible. Doing this is no problem. Just whack in a couple of willow stakes, which are almost sure to take root and provide a couple more little trees, and attach a cross piece. You can use better quality timber for this if you wish, but it seems an unnecessary expense when all the materials are close at hand and free of charge.

I'm not qualified to write about more elaborate bridges across the main river. The height of my ambition is a bridge made from a couple of telegraph poles, with slats of wood nailed to them. Unfortunately, like railway sleepers, telegraph poles are not too easy to come by. In my area there is a waiting list for them. If they can be obtained, they do a super job and last for years. Where there is no existing bridge, permission will

Fig 67   Basic pole bridge plus handrail for greater safety.
This bridge was improved by the water authority, so that the
weed cutting boat could pass underneath.

almost certainly be needed to build one. Apart from the desirability of keeping the bridge out of the reach of flood water, it may also be necessary to build it high enough above water level to allow the weed-cutting boat to pass underneath. On one of our stretches of river, there are two of these telegraph pole bridges, quite close to each other. They had been made too close to the water to allow the boat to pass underneath and it used to take the men longer to haul the boat in and out of the water than it did to cut the weed through the whole length, a distance of about two kilometres. At last common sense prevailed and bridges were raised, at least one end was. For the past six months, the two bridges have remained at an uncomfortable angle, but I suppose the job will be completed eventually.

There is a much more sophisticated bridge at the downstream end of our stretch. This was built by a member who has clearly missed his vocation, as he should have been a civil engineer. He cannibalised an old trailer and, helped by his fitter friend, prefabricated the bridge at work and brought it, in two halves, to the river where he completed the assembly. It has stood for seven years, as good as new, except that it now needs a fresh coat of paint. A job like this would cost a great deal of money if an outside firm had to be paid to construct it but, you never know, you may have a member who has both the necessary skill and access to suitable materials. Fig 96 (page 150) shows what a splendid bridge it is.

SEATS

Access to the whole fishery, river or lake, should now be perfectly straightforward and a few suggestions about the fisherman's creature comforts might not be misplaced. Most anglers, especially when fishing their home water, do not want to flog away all day. An 'away fixture' is a different matter, as the chance to fish that particular water may not come round again very soon. A fishery is a much more agreeable place if a few seats are set up around it or along it, as the case may be. All that is needed is a couple of uprights dug into the ground, and a cross-piece nailed to the uprights, to make the seat. Nothing could be simpler, except when there are cows around! The plain fact is that, if you have cows, you can't have seats. A major feat of construction is needed to make a seat strong enough to withstand even a gentle shove from a cow. But even without cows, it pays to build seats to last.

I am always on the lookout for free offerings which may come in useful. A few years ago, the council was replacing a wooden footbridge and some very nice creosoted lengths of wood were lying around, just right for the seat part, rather over a metre long and at least four centimetres thick (3 feet by 1½ inches). At the same time I was lucky enough to come by some offcuts from railway sleepers, just long enough to make the uprights for the seats. So, there was a nice supply of fisherman's seats, free of charge, except for a few long nails to fasten the seat to the uprights and a bit of hard labour to dig the uprights into the ground. They need to be dug well in, otherwise the end product will be a pretty wobbly seat.

The question now arises of where to site them; obviously somewhere pleasantly secluded. The seat also needs to command a good view of the water, so that signs of activity can be spotted and, if possible, it should

Fig 68   Bridge made using two wooden pallets, supported by stout poles.

Fig 69   This seat is made using off-cuts from old railway sleepers.

be reasonably sheltered from the wind, so that members can have a picnic in comfort. By the lake, it is quite useful to have a few seats placed so that members can fish from them. It makes fishing much easier for the older members and it also suits those who have a more relaxed attitude to our sport.

THE FISHING LODGE

The final amenity to be discussed is a fishing lodge. This edifice typifies most strongly the difference between river and lake fisheries. Let no one accuse me of saying that river fishermen are unsociable; a couple of the most unsociable anglers I have ever met were lake fishermen, but fishing the river is a more solitary sport. As a general rule, the larger the club, the more need there is for a fishing lodge. It follows, therefore, that as a stretch of river seldom has more than three or four anglers present at the same time, there is no real need for any such amenity. Where the rules require members to book in before starting to fish and to make a return of their catch on leaving, somewhere has to be found to keep the record book. One of our river fisheries has a little garden shed and each member has a key to the padlock; another club just keeps the book in one of the farm buildings; two others, I'm ashamed to say, seem to have a very cavalier approach to statistics and don't have a book at all! The merit of all these arrangements is that they entail very little expense.

I am lucky enough to visit one river fishery which stands right at the other end of the amenity spectrum. The club's meeting place is a room in a beautiful old house, overlooking what used to be the mill pool. It is like going back to another age just to sit in one of the old leather armchairs, looking out through the big bay window and watching the trout rise. The standard of fishing and keepering is on a par with the amenities. To fish there is a rare treat and, like most of us, my normal fishing is more mundane, the fishery itself needing much hard work and maintenance if reasonable sport is to be enjoyed. On rivers, however, there is no *real* need for a fishing lodge.

On the lake, some sort of fishing lodge is almost essential, especially if any kind of community spirit is to be built up. A fairly large lake will often have twenty or more members present at the same time and lake fishing can be a pretty tedious business on occasions. There are those anglers whose temperament and tenacity enable them to fish on regardless, but other less dedicated souls may be quite content to shelter from the elements, drink a cup of tea and have a chat, while waiting for conditions to improve and to see some sign of a rise. At the end of the day, too, there is time to chat and admire each other's fish. In my view, the lodge should be as large as the members can afford and it is essential that adequate toilet facilities should be provided.

Once the basic lodge has been built, improvements will come steadily. The biggest advantage of all will be the installation of electricity. If the situation of the lake is too remote, this may be almost impossible on the grounds of cost but, if the job is just reasonably possible, it should be undertaken at the earliest moment. It's not simply a matter of changing from a chemical to a flush toilet, nor merely of electric light, cooking facilities and a fridge; the fish-rearing potential will be greatly enhanced, as electric pumps and aerators can be installed to reduce risks during spells of hot weather and low oxygen. Once the basic work has been done to the lodge, it is surprising how improvements will occur almost unnoticed. One member will produce a set of mugs, another some

Fig 70 The Lodge. The scales and fridge for fish are out on the verandah to discourage members from bringing their slimy fish indoors.

*Fig 71   A lakeside seat. 'Uncle' George is almost ninety and
well entitled to sit down on the job.*

cutlery, a few easy chairs will appear, the supply of sugar and teabags
will be topped up. As I suggested earlier, the lodge does much more than
provide a bit of shelter and comfort – it has a marvellous influence in
fostering a good club spirit. A hidden advantage is that, when the
fishermen are busy talking and drinking tea, they will not be making
serious inroads into the trout population, thereby decreasing the need for
restocking!

My aim is to produce a fishery which is in harmony with its
surroundings. Lowland fisheries are almost invariably situated in areas
which are intensively farmed; agriculture often comes right to the water's
edge, so the improvements and amenities I have suggested are a natural
part of this man-made environment. When I go north, I certainly don't
expect to find a flush toilet by the side of some remote loch, nor seats to
rest on after scrambling across the rough ground. Such amenities would
not only be out of place in this setting, they would actually detract from
the pleasure. The fishery manager must have an overall picture of what

he wishes to achieve in his mind, but the standard to which a lowland river or lake is maintained depends ultimately on the wishes of the members and the amount of work they are prepared to do themselves, or to pay for. I like to think that the waters for which I have some responsibility offer the members a relaxing day out, in pleasant surroundings, and with a fish or two in the bag as a bonus. We are certainly not catering for the fanatics and the fish-hogs.

6
Stocking

If you were running a coarse fishery and had done your work well, the probability is that you would need to *de-stock* your water, in order to avoid the evils of over-population and poor growth-rate. Trout, however, tend to suffer from severe headache on being caught. Some fisheries insist that all fish caught are killed, others allow catch and release. My fisheries fall into both camps; all fish caught in the lake fisheries must be killed, whereas, in the rivers, we allow fish to be returned. I have a fairly open mind about catch and release. I feel that if a fish is to be returned, the angler should bring it in as quickly as possible, risking the fish breaking free; the fish should certainly not be played out. Provided that the fish has not been badly hooked – in which case it should always be killed – it can then be liberated, preferably without even taking it from the water. Unfortunately, some anglers are very clumsy, and any fish they hook, even if they are using barbless hooks, has precious little chance of survival.

Whichever practice you favour, re-stocking is still going to be essential; however much work you may have done to improve the spawning potential of your river, the demand for fish is almost certain to exceed the natural supply. In lakes, of course, even if you do practise catch and release, rainbows are short-lived, so fresh fish are bound to be needed every year. The aim of your stocking policy should be to provide good fishing – whatever that may mean. Some of my ideas on the subject will hopefully become apparent in the course of this chapter.

When considering your stocking policy, a cautionary tale might not come amiss. Many of you will know Skues' story of Mr Castwell, who died and went to 'another place'. He found himself by a delectable river, accompanied by a faithful attendant. He was delighted to catch a beautiful 2lb brownie first cast and thought he must be in heaven, but when he discovered that he had to go on catching beautiful 2lb brownies, from the same spot, for ever, he realised where he was. Big fish are super, but if *all* your fish are big, then each one is only average and the extra pleasure that the big fish gives has been lost. As with improving the habitat for your fish, variety is the spice of life and a stocking policy should aim to give as much variety as is reasonably possible, so that the angler never knows what he is likely to catch next. Though our policy is to stock only brown trout in our rivers, every effort is made to ensure a wide variety in size. Only recently, the fish farmer contacted me about our next order and, out of 300 brownies, we agreed on 100 11–12in fish, 150 averaging 1¼lb, and 50 larger ones. The angler may occasionally grumble about the smaller ones, but such a policy makes him appreciate the larger ones all the more.

On the lakes, we use both brown trout and rainbows. We had a brief flirtation with Brook trout, but in our water they were a failure. They were far too easy to catch when first stocked, but then the survivors seemed to disappear completely. In our largest lake, which is over 25 acres in extent, we can give an even greater variety in the size of fish stocked. Only a few weekends ago I was taking my usual active part in fishing, sitting in the hut drinking tea and talking to a couple of other members. My wife appeared, looking more than usually self-satisfied, and plonked a 6¼lb rainbow on the scales. This galvanised me into activity and I promptly caught a brace of fully a pound apiece! Part of the charm of fishing is its uncertainty.

Nor should you make the mistake of thinking that quantity makes for good fishing. A few years ago I went to stay with a friend in Wiltshire and, hoping to give me a special treat, he had booked us a day's fishing on a sidestream of the Avon. He had been told that the stream held some very good trout – which it did – but what he had not been told was that the trout farm had just lost about 20,000 12in rainbows down this stream. By mid-afternoon, we agreed that we were tired of catching fish and decided to call it a day. It had been fun, but with the hundred up, enough was enough! There is certainly a lot more to fishing than just catching fish. This may be an extreme case, but some fisheries, especially commercial ones where the customer has to pay a fixed price per pound for the fish he catches, tend to over-stock and to make the fishing far too easy.

The stocking I have just mentioned was an accidental one, but you should be warned by the sad tale of over-stocking on the River Bourne, as told in *Where the Bright Waters Meet*, one of the finest of all fishing books (*see* Further Reading). In this case, some members of the little club felt that the river could be improved by re-stocking. Far too many fish were put in to the river and the beautiful native trout starved to death. It took the river many years to recover from this disaster. If you have doubts, you should certainly seek advice from your local fisheries officer or even your supplier. Better too few fish than too many, though anglers will make swift inroads into the numbers. Money – or the lack of it – also exercises a powerful restraining influence. This year, I have overspent on my first two stockings and members will have to be satisfied with only 12in fish for the final stocking. Fortunately there are still plenty of the larger ones left from previous stockings to provide the variety that I keep insisting on.

During the initial years of running a fishery, I think that money for fish is better spent on providing trout of takeable size rather than on buying smaller fish and hoping to build up a stock. Do not, however, put in fish which are appreciably larger than your water could reasonably support. In the upper reaches of the little river which I look after, a trout of a pound is more than big enough but, farther downstream, trout of twice that weight are not unreasonable. I was amazed when helping the eel trapper, on the downstream reaches, to see how many caddis came up with his nets. The boat seemed to be full of bits of twigs, but twigs which crawled round all over the place. Quite a good supply of crayfish came up in the nets as well, so I'm not too worried about our trout's menu.

How often you should stock again depends on the amount of money you have available and the fishing pressure on your water. In our big lake, where we have over a hundred members, we stock almost every weekend until at least mid-July. Apart from replacing the trout which have been killed, we find that an injection of new fish seems to stir up the other ones as well and this helps to maintain quite a lively standard of fishing. In the rivers we don't need as many trout because the bag limit is small – two or three fish, depending on the fishery – and catch and release is normal. Even so, we like to have at least three stockings during the season. In the first place, it is really most unwise to put all your eggs into one basket: a sudden flood can sweep almost all your stock fish away. The trout have come from the farm, usually living in still water, and they don't cope very well with extreme conditions. By staggering the stockings, the water should never be overcrowded with fish. A single stocking often leads to hectic, but far too easy fishing during the first weeks, only to be followed by a rather dull end to the season. I find that three stockings give us fairly consistent sport, with only a few days where the fishing is too easy. Finally, a series of stockings avoids making an excessive demand on the available food supply. The last stocking is

Fig 72   No shortage of members to help with the stocking.

usually made at the end of July, when all weed-cutting is finished, so that the final batch of fish will not have to suffer any extra disturbance.

When stocking, the distribution of the fish is important. In the wild, brown trout are solitary creatures, each having its own territory and guarding it. Rainbows seem to be a bit more gregarious. Both species, however, have been living in the unnatural conditions of the fish farm, where even the brownies live crowded together. For at least a little while after stocking the brownies still tend to stick together, if you put them in together. I have no doubt at all that it pays to spread your stock fish out as widely as you can, especially when putting brownies into the river. If the whole bank is easily accessible by car, the supplier can drive his vehicle along, putting the fish in at intervals. On many fisheries, the banks are much more difficult: only the upstream part of one of my fisheries can be stocked in this way. The rest of the fish are then put into a floating cage which can be towed behind the punt. One person rows while his companion slips the trout in, one or two at a time, with his landing net. I think this is the best way of all for stocking the river and there is never any trouble finding members to help with this job.

*Fig 73   Cliff, with a brace or two that he wouldn't mind
catching himself at a later date.*

Even with rainbows in a lake, it is better to spread them out as far as possible, if only to stop the odd greedy member standing over a shoal of fish until he has caught his limit. I fear that in any club, especially a large one, there are always a few members for whom catching the fish is the only thing that matters. One just has to accept that different people get their enjoyment from fishing in different ways; some suffer from Limits Disease, others even go in for competitions, as if catching fish wasn't hard enough already! Tastes differ. I had all the fun I wanted one evening recently, failing to catch anything at all. I was trying to catch one of the big brownies which cruise up and down near the bank. I did finally hook one of about 3lb; it escaped, but I was well satisfied.

MAKING A FISH STOP

One problem normally encountered only on a river fishery, which can upset the river manager, is that the trout can actually 'vote with their feet', or tails, and leave! Unless you have a very long stretch of river at your disposal, it is wise to have a fish stop somewhere near your bottom boundary. Trout coming in from the farm are not accustomed to running water and, during the first few days after stocking, they tend to drift downstream and quite a lot of them may be lost to the fishery. If there is a grid across the river, many of the drifters will work their way back upstream until they find somewhere to live.

You will not need, or want, to have a grid in place for more than five or six days. It requires cleaning at least twice a day, to ensure that the water can flow freely. If you allow it to clog up, you will have made a weir and flooded the farmer's fields; this is likely to make you unpopular. If it doesn't make a weir, either the whole lot will collapse under the pressure of water or the river will scour a hole in the bank. Neither of these is a very attractive proposition, so it is vital to keep the grids clear. Some trout will still migrate. I know that a few of mine drift off downstream and give the youngsters, expecting to catch a few eels, a pleasant surprise. Even so, the use of a grid probably ensures that the majority of trout stocked stay within the fishery.

Choosing a Site

Selecting a suitable site for the fish stop is of paramount importance; mistakes here mean loss of fish and, worse still, a lot of hard work which has to be done all over again. The stop needs to be situated in a straight stretch of river with firm banks on both sides. We neglected this, to our cost, on one fishery; the banks seemed firm enough, but one side was actually silt which had been deposited by the river. The site chosen was also on a bit of a bend. One night the grid blocked and the river promptly cut itself a new channel, through the old silt bank, by-passing the grid completely – some fish stop! Nor should the fish stop be sited on a shallow. Another club I know of tried this and, even on quite hard gravel, the river soon scoured out quite a deep hole. Select a site where the river bed is as firm and level as you can find and in which you can work quite comfortably in chest waders. All being well, this will be about the average depth of your river. If so, you will encounter fewer problems with water pressure and scouring. The greater the area of grid through which the water is flowing, the safer your work will be.

Construction

Initially, making a fish stop is the same as constructing a pair of groynes, though you must be very careful to use wire netting deep enough to be weighted down to the river bed with some heavy stones, flints or sacks of earth. Having made a strong groyne on each bank, the gap in between can then be fitted with grids each time the river is stocked. If you have to have grids made for you by an outside firm they are very expensive, but the gap between the groynes should be as wide as possible; certainly not less than two-thirds of the width of the river. When we needed grids for one river, the school at which I was at that time teaching was being re-wired and I was lucky enough to get a load of old conduit, in pieces long enough to make the bars of a grid. The farm foreman made me a very good grid, if a bit heavy, by welding the conduit to a frame made from angle irons. I was also lucky enough to borrow (on a very permanent basis) quite a long grid from another club, so on one fishery, at least, bridging the gap was a very cheap job. Another club I know of had a member who owned an agricultural engineering business and he made their grids – very sophisticated ones, too – at a very advantageous price.

However you acquire your grids (as a last resort you may even have to pay for them) they need to be strong. People seldom realise the terrific pressure which even a small river can exert. Fig 74 shows what happens when the bars of a grid are too flimsy. As you need strong grids, they are therefore likely to be quite heavy, so it is better to have several small grids which can be put in place side by side, rather than one monstrosity

Fig 74 The grid must be strong enough – the one on the left wasn't! The right-hand one is made from scrap conduit, welded on to angle irons.

which no one can lift. Don't be tempted to save money by leaving a gap between the groynes which is too narrow – that is false economy. If the grids clog and the river floods or scours out a new channel you will have wasted all your time and money. Incidentally, if there has been heavy rain and there is a risk of a flood, take the grids out immediately; better to lose a load of fish than risk unnecessary damage.

The most satisfactory of all my fish stops had one very worthwhile refinement. After building the two groynes, the metal bar, which was actually three angle irons welded together, was rested on the river bed and the wire netting was stretched back upstream, protecting the river bed and reducing the risk of scouring. The netting had then to be held firmly to the river bed. I used a load of flints to do this, as I had a pile of

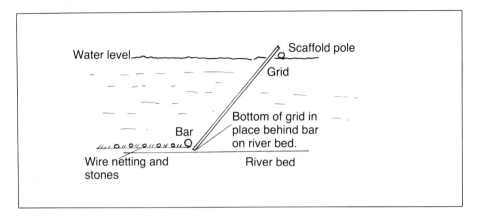

Fig 75 Side view of the preparation of the river bed, ready for making fish stops.

them available. Plastic sacks filled with earth or gravel would do equally well and could make a much neater job – don't forget to punch a few holes in them to let the air escape. The grid is then dropped in immediately downstream of the metal bar, making an effective and fish-proof joint.

It can be seen from Fig 76 that the grids should not stand vertically, but slope appreciably downstream, so that weed and debris brought

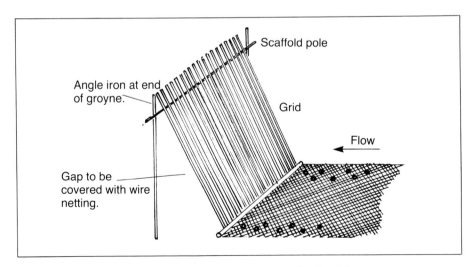

Fig 76 The fish stop, showing angle at which grids should be placed to reduce clogging of fish stop.

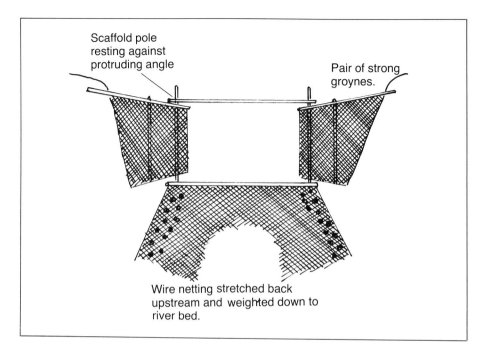

Fig 77   Construction of a fish stop.

down by the current will tend to rise to the surface. The tops of the grids need to rest against something. The simplest thing to do is to leave about 15cm (6in) of angle iron protruding at the river end of each groyne. A long scaffold pole will then bridge the gap between; it is easy to put in place each time the grids are needed, and equally easy to remove, so that the channel is not blocked in any way. I have used wooden poles to do this job but, unless they are quite big trees (and therefore weigh a ton), they are not really strong enough for the job. Because the grids are not vertical, there will be a gap between groyne and grid. I just block that with a bit of wire netting, which seems quite satisfactory; the gap is quite small and the pressure of the water holds the netting firmly in place.

As stated earlier, the grid should be cleaned morning and evening to make sure that it does not clog up. It is good policy to stretch some wire netting right across the river, about 10 metres (11 yards) upstream of the grid. The netting need only extend 30–40cm (12–16in) below the surface, but this will act as a first line of defence and catch a surprising amount of rubbish which would otherwise clog up the grid. People upstream find that the river is an excellent receptacle for their kitchen

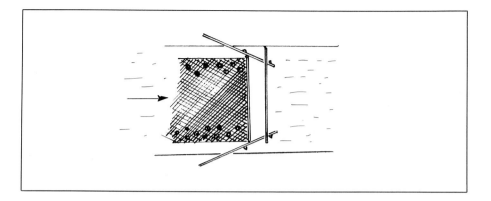

Fig 78   Aerial view of fish stop.

and garden waste: cabbages, lettuces, carrots, sprout stalks, hedge and grass cuttings all float down the river in abundance! The best tool I have found for cleaning the grids is a lawn rake. The tines slip nicely between the bars of the grid and the rubbish can be lifted out easily.

OVERWINTERING

Despite all the good work which has been done to make your water more attractive to the trout, almost all the stock fish will have vanished before the start of the next season. I have watched stock fish spawning in December but, after that, they disappear never to be seen again. In rivers, this seems to be a fairly normal state of affairs, especially when the stock fish are brought in from a completely different water. Apparently, this is a common problem. I was reading an article quite recently in which the writer was complaining of the same thing, although he was living in another part of the country from mine. What I don't understand at all, however, is where the fish go. One rarely sees a dead trout, and trout are very seldom caught more than two or three kilometres downstream of our stretch; even the youngsters fishing quite close don't catch them in significant numbers. We know, not only from our catch returns but also from actually seeing the fish, that a good stock of fish remains at the end of the season, only to vanish into thin air, or deeper water, before the next season. In lakes, of course, there is more carry-over of the previous year's stock, though even here, much depends on how well the fish have settled in their new environment. When considering our stocking policy,

Fig 79 It would be hard
to afford to buy in trout of
this quality.

Fig 80 A bin-full goes
straight into the lake.

we always assume that we are starting from scratch each year and that any overwintered fish are a bonus.

REARING TROUT

The first part of this chapter has assumed that the fishery is buying in fish from a farm for immediate stocking and capture. It seems inevitable that most trout fisheries must depend, at least to some extent, on put and take. Rearing your own trout is a much more attractive proposition if you have the necessary facilities and are prepared to take the risk. You will have losses, possibly lots of them to start off with, until you gain the necessary experience and expertise. You should certainly also keep money available to buy in fish if you need to. I wouldn't presume to write about fish rearing, though one of my fisheries does it most successfully, buying in fish at 4–5in long and producing trout of up to 7–8lb. For advice in this field you must go to the professionals, but I will suggest a few of the essential requirements if you are thinking of rearing your own fish.

Before you even go into the planning stage, you must realise that someone has to be available every day of the year, not just to feed the fish (that can be done automatically) but also to check on their well-being. Screens must be kept clear of weed so that the water flows freely; the trout need to be checked for signs of stress or disease; holding pens need to be cleaned out when empty. Unless someone is available every day, therefore, the operation is not really worth considering. Another factor to take into account is the number of fish you require – too few, and it's hardly worth the effort and worry. I think that you must be running a fairly large fishery, needing several thousand fish a year, to make the effort and expense worthwhile. On the fishery where we rear our own fish, we often have as many as 10,000 trout on the site at the same time, though over half of these will be the small fish growing on for next year's stockings.

Having decided that the effort might be worthwhile, then start to see how practical the scheme will be. To have a really good chance of success, the stock ponds need fresh water running through them all the time, cool, well oxygenated water. You will notice how often fish farms are situated where spring water is available or where there is a borehole producing water of suitable quality. In this way the farmer has a supply of pure water, and in summer water which is relatively cool; high summer temperatures are a recipe for disaster. On the river, for example, a site may be excavated just downstream of a weir with an inflow pipe taking water from the higher level and running it through the pond or ponds which have been made. The sides of the ponds need to be built up all round with the spoil from the excavation to give as much protection as possible against flooding. A friend's club rears its trout in this way, producing fine brown trout to over 4lb. Their enemy is a heatwave, something that doesn't happen too often but, if danger threatens, they have to mount a rescue operation and tip into the river all the fish which are left after the early season stockings.

On one of our lake fisheries, there is a river running round the west and north banks. There is quite a large fall from river to lake, especially in summer when the river is penned and the level rises. A 12in pipe has been laid which, when fully open, pours a terrific volume of water into the lake, freshening up the whole fishery and producing a good flow through the stock ponds, where the water flows out of the lake. The lake is also spring-fed, so it was always possible to rear some trout in the outflow, but bringing the water in from the river has more than doubled our fish-rearing capacity and considerably reduced our problems in summer. There are two outflow channels from the lake, both of which feed long and narrow stock ponds in which the smaller fish are grown on.

Once the fish reach about a pound in weight we begin to stock them into the lake, although by July many of the better growers will be in excess of 2lb. We normally keep about 1,500 of these fish to grow on for a further year. These are transferred to a bay in the main lake which we have netted off to prevent random stocking, though we have had occasional escapes. By the following season, many of these fish will be over 5lb in weight. Netting the bay is by no means one hundred per cent successful. Every year, some fish are missed which, if they survive, provide a few real monsters the following year. The bay is between 3 and 4 metres (10 and 13 feet) deep, so the fish remain cool in hot weather; they have plenty of room – free-range fish, in fact – so that we produce not only big but also beautiful fish. The brown trout, in particular, are superb and I think our supplier is quite envious when he sees the end product.

There are two main advantages in rearing your own fish, provided that you are reasonably successful. The first advantage is one of cost. It would have to be a very wealthy club indeed to afford to stock the large number of big fish that we do, even if you could find a satisfactory source of supply. The biggest advantage, however, is that you are stocking trout which are almost native fish, having been raised since they were very small in the same water into which they will be released. We are convinced that fish like these adapt well when stocked as compared with trout stocked directly from the trout farm. Certainly the winter pike-fishing weekends produce more trout than pike, some of them glorious fish in splendid condition, which are clearly not going to waste their time spawning. The few pike that survive aren't bad either, having a lavish menu and little competition to eat it; the best so far weighed 39½lb!

I am equally convinced that the same advantage applies to rivers. When I visit my friend's club, where they rear their own trout, our winter coarse fishing is quite often disturbed by large, but unwanted, trout. This stretch of river has little natural spawning and these are all stock fish from earlier years which have settled in well. We don't rear our own trout on the stretch of river that I manage, but there is a lake fishery higher up the valley, with the stream running alongside, and the owner is currently rearing some brown trout in the river water. We will buy some of these in the hope that they will adapt better than the trout which have come in from different water. Many famous river fisheries produce their own fish and I am sure that it is not only the cost advantage which is so attractive.

The disadvantages of rearing your own fish are obvious and you always need to keep some money in the kitty so that you can buy in stock fish if (or when) you have a disaster. We have been hit at various times by eye-fluke, liver disease, high water temperature and floods. Usually the fishery scientist will tell you what has happened though, on one occasion, the comment was that they were absolutely super fish, and the only thing wrong with them was that they were dead! The more you do the job, the more your standard of husbandry should improve, but you should always pay attention to basic cleanliness. Small rainbows seem to be very subject to eye-fluke. When our ponds are empty of fish, we poison them with the dosage prescribed by the scientist to kill the snails which harbour the flukes. Once or twice a year, the flow through the ponds is shut off and malachite green (available from chemists and the fishing trade) is put into the water. This helps to control fungus infections. As I said earlier, I am in no way qualified to write about fish

Fig 81   Charlie puts some
trout in the boat.

Fig 82   Nick sets sail to distribute
the fish as widely as possible.

rearing. I have simply given a few indications of what our experience has been. Before you embark on any such venture, seek help and advice from your local fisheries scientist and from your supplier. Setting up the project is expensive and you need to make sure that you have a good chance of success.

Whether you buy takeable fish from a farm or rear your own trout, your aim is still to provide 'good fishing', which brings me back to the question posed at the beginning of this chapter: what is this elusive thing called 'good fishing'? It is certainly not to be confused with easy fishing; too many fish and the sport soon becomes boring and repetitive, and the fisherman loses interest. This is bad fishing. To move to the opposite extreme, the angler must always feel that he has a chance of catching something. Fishing when the angler feels that there is no chance of success and all hope has been lost, is just as bad as fishing which is too easy. The fishery manager is seeking to create a happy medium but, alas, even this is not as simple as it sounds. Good, bad and indifferent fishermen, we all *enjoy* our sport. The problem is that, what is quite easy fishing for some anglers, is jolly hard for others. It is an established fact that a few anglers catch the bulk of the fish. In a lake it is almost impossible to resolve this problem. There must be enough fish to allow the average angler to catch something so, almost inevitably, the fishing will be a bit too easy for the most skilful. Life is a little easier on the river. Some parts will allow comfortable fishing and casting but other parts will be much more awkward and the expert can find a real challenge. To some extent, I have over-simplified the problem of good fishing but, as long as you don't confuse quantity with quality, you will be on your way to success with your stocking policy. Remember, too, what has been said about variety; the uncertainty of the sport is another ingredient of good fishing. The angler should never know what is coming next – mystery is perhaps the greatest element of all in good fishing.

7
Predators

The fishery manager has much the same sort of job to do as a gamekeeper in that he must control the number of predators and competitors in his area as far as possible. It is not a simple job. A stretch of water can only provide good feeding for a certain number of fish, but there is little point in killing all the pike if the trout's other competitors flourish to such an extent that the trout go hungry. A gamekeeper doesn't have quite the same competition problem, as he goes round his land keeping the pheasants well fed. I don't really see that you can go round bombing the trout with pellets during the fishing season, though one keeper I know has a very good system. He lives in a lovely house overlooking the mill pool and always makes sure that his fish are well fed once the season ends. This discourages the fish from straying, helps to keep them fit and growing and brings them quickly back into good condition after spawning. The crafty devil has one ultimate refinement; he is a super fisherman and he catches them from time to time on nymph or dry-fly, and puts them back sadder but wiser fish. The end product of this good management is a fine population of visible, large, well-educated trout. The club members can spend hours failing to catch them, but be highly delighted to see such fine fish available for capture.

PIKE

Of all the predators, pike is almost certainly enemy number one. The little ones will eat the little trout if you are lucky enough to have some natural spawning, and any pike from about 3lb upwards is a big threat to your stock fish. The first time that you see one of your newly-stocked trout swimming downstream sideways, you are likely to be a little surprised, but closer inspection will almost certainly reveal that it is firmly clamped in the jaws of a pike.

The pike menace is probably more easy to control in rivers than in lakes because, unless the river is too big, electro-fishing can be carried out very successfully. Yet again, a good relationship with your fishery officer is an enormous advantage. Our water authority men are pleased to pay us a visit every two or three years, and when they come they do a pike cull throughout the whole stretch; if we have a surplus of coarse fish, they take away some of them as well, to use for their restocking. One length of about 300 metres (330 yards) will be fished very carefully two or three times, to provide a population census, so that changes can be monitored. However, electro-fishing can be a dangerous business and should only be carried out by those who have been properly trained. Permission is always needed for this type of work and, in practice, it is highly desirable that the authority's men do it themselves.

The other really big ally you can have in your unending war against pike is the eel trapper. He can work successfully in both rivers and lakes

Fig 83   In the river pike like the slack water behind a groyne.

– though the lake will have to support an eel population large enough to make it worth his while coming. Our local trappers use fyke nets, which look like enormous keep nets, and the eels and fish find their way in through a funnel, but can't find their way out. Our trapper visits us twice a year and, though we don't have a big pike population, he almost always kills half a dozen or so during his week's work.

Some keepers snare pike very successfully, though I have not tried this myself; others shoot them, if they are lying in sufficiently shallow water. A friend made me a 'trident' (*see* Fig 6, page 18) and I mount this on an aluminium television aerial. It is an evil weapon, has quite a long reach and has accounted for several fish, especially ones lying too deep in the water to shoot. Its biggest victim so far was a pike of 10¼lb. We had been trying to get her for several weeks and when finally killed we found a nick near the tail where we had just missed her on an earlier occasion. If you think you have a pike problem, it is hardly likely to match that encountered in some of the unpolluted waters of fifty or sixty years ago. Writing early in the 1930s, E.C. Keith describes the removal of well over a thousand pike from a stretch of the river Wensum, in Norfolk. Even with this number taken out, he decided to leave plenty of small roach to help feed the survivors, lest they eat all his trout! These fish were removed by netting, never a very easy operation, especially where the bed is uneven. Gravel pits can be very hard indeed to net, with their sudden changes from deep to shallow water. We have netted one of our gravel pits, both for pike and coarse fish, but failures far exceeded the successes and we no longer bother.

You should not discount angling as a means of reducing the pike population, especially in still waters where control is more difficult. During the trout fishing season, lure fishermen make heavy inroads into the population of small pike and occasionally the not so small. It has not been unknown for a 20 pounder to succumb to the charms of a muddler minnow! Once trout fishing ends, some organised pike fishing can pay handsome dividends. On one of our gravel pits, it was almost worthwhile being a member just for the pike fishing during the early years of the club's existence. One memorable New Year's Day yielded well over a hundredweight of pike in a few hours from one quite small bay. Now, however, pike are much scarcer. Some pike-fishing weekends scarcely produce a single fish, but for the specimen hunter, what a fish that single pike may be, having lived with little competition, on a lavish banquet of our trout.

Fish traps can also be used to make inroads into the pike population

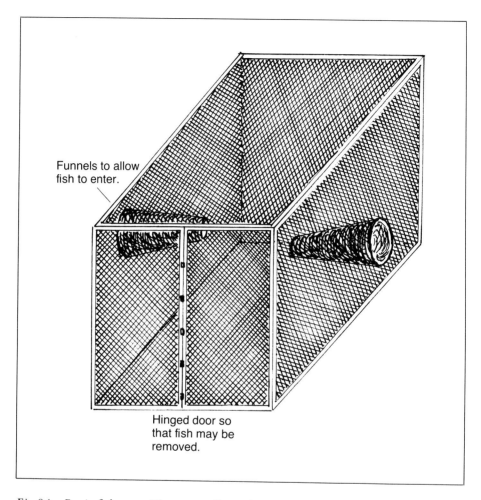

Funnels to allow
fish to enter.

Hinged door so
that fish may be
removed.

*Fig 84   Basic fish trap. The trap will need some weight
inside to sink it, and a rope attached so that it can be pulled
out.*

and, of course, to reduce the number of other competing fish as well. At its most simple, a fish trap is just a very large box; all the sides are made of wire netting and one or two of the sides will have a funnel let into them. The fish are curious, swim round the trap, go in through the funnel and cannot find their way back out. A rope is attached to the frame, so that the trap can be pulled ashore and the fish taken out. In general we found that the trap worked most effectively from May until the end of the summer.

One club made a large fish trap on their lake and sited it alongside a bridge across to an island. The trap stretched out from the shore for about four metres and the rest of the gap to the island was netted off, so that all fish swimming that way were likely to move along the side of the trap and many of them find their way in through the funnels. Some very big catches of fish were made at times. On one occasion, there was a shoal of bream, some of which were so big that three of them were quite enough to fill one of the plastic dustbins used during trout stocking – one of the largest bream weighed just 10lb. As other lakes on the site are available for coarse fishing, the bream were not wasted. Now, twenty years on, the trap still catches pike and a fair quota of nice tench but very few roach and bream. It seems that trout are now the dominant species, but the fishing effort is not relaxed, in the hope that the population of unwanted fish will remain relatively small.

EELS

Next to pike, I regard an excessive eel population as the greatest curse. They are ferocious predators; any moderate-sized eel will almost always contain the remains of small fish. Some, alas, will contain the remains of crayfish, a source of food which I would much rather find inside the trout. I am also a keen coarse fisherman, and eels play havoc with the spawn of coarse fish. I have watched them in the clear water of the lake, at spawning time, gathering for the feast. Some rivers have eel traps, built at weirs and mills, to capture the eels as they migrate but, except for a bonus clear out when we are doing some electro-fishing, I have to rely on the eel trapper and his nets to keep our population within acceptable bounds. He comes twice a year, first of all in late spring when the eels are on the move again after what seems to be a period of hibernation, and then again in September or October when the eels are migrating. He catches enough eels to make the venture profitable and the fishery also profits; fewer pike to eat the trout and more food for the trout to eat. The eel's only redeeming feature is that it is so good to eat. I would rather have eel than trout on my menu! If you have any doubt about the evils of the eel, our trout supplier has no illusions. He put several thousand fingerlings into a concrete holding pond. After a while it seemed to him that their numbers were dwindling sharply. The pond was drained, and three large and very well-fed eels were found hiding under a sill.

PERCH

On some fisheries, especially those in still waters, perch can be a nuisance; they certainly eat some of the food which would otherwise be consumed by the trout and they will take a lure very readily. A couple of my more misguided friends have been quite happy some days to catch perch when fishing at Chew Valley, great big fish running to well over 2lb apiece. That makes expensive coarse fishing, especially if you have taken a boat, but the quality at Chew is rather exceptional. Unfortunately, perch are very scarce in many waters at the moment, as they don't seem to have recovered from the disease which all but wiped them out about ten years ago. Before this, we had an excessive perch population and we found that we could at least keep it in check by putting bundles of sticks into the water at spawning time. The perch will

Fig 85   Brownies this size will eat anything, so don't stock with small trout.

spawn on these and the bundles can then be removed. Netting, too, provides a useful place for coarse fish to spawn on, especially cyprinids. A couple of years ago, a team from one of the universities left some eel nets in a lake over the weekend, and then had to leave them in much longer than they bargained for; the roach and bream had spawned all over them. A much more sophisticated form of this is now one of the techniques used in fish farming. Fish can be induced to spawn on small sections of netting and the spawn moved to where it is wanted.

BIRDS

Our feathered friends also feature on the list of predators. Herons are protected, but unless they are too numerous, I don't worry very much about their depredations as they are very much a normal part of the waterside. Obviously stock ponds have to be carefully protected (ours are completely netted over), but I doubt whether herons take an unacceptable number of trout from rivers and lakes. As a general rule I'm no great exterminator of predators and competitors, I just want to leave a reasonable balance in favour of the trout, but there is one bird for which I have no time at all – the cormorant. It is fully protected but takes a terrific toll of fish at a time when fish stocks in most waters are already under great strain. As for the place they select for their roost, it becomes just a filthy polluted area of dead trees, lavishly whitewashed with excrement. Before they were protected, we did all we could to discourage their unwelcome visits. One Christmas the keeper gave several of them a present of lead shot. One of the dead birds looked a bit odd so he gave it a shake and a 10in carp fell out, still alive and kicking. It swam off seemingly none the worse for its experience, and should be big enough now to be well worth catching. Another day, two birds disgorged over twenty small roach which had only just been swallowed; I put the fish in my deep freeze to use as deadbaits. On yet another occasion, on a river in Yorkshire, the keeper and I found a cormorant swimming in the river so gorged with fish that it couldn't take off. Cormorants are not my favourite bird, and it's a pity they don't stick to the sea coast where they belong.

Fig 86   The President, admiring his catch.

ANIMALS

In Norfolk, we don't have too much of a problem with animal predators. Mink have not yet made their unwelcome appearance and otters are unfortunately almost extinct. We did find a pile of crayfish shells one day, which suggested that an otter might have been about, and we certainly wouldn't begrudge one a meal of our trout. Our only real animal predator is the poacher.

POACHERS

In general, the poacher is more of an irritant than anything else. Our poachers are usually local boys, during the school holidays, and I begrudge them every trout they steal, especially as we have made about half a mile of water available to them, free of charge. They usually see me coming a mile off and any reasonably healthy lad will run a lot faster than I can. If you do manage to catch one, it helps if you are also a water authority bailiff, armed with a proper warrant; you can be almost certain that the offender will not have a licence, so at least you can report him for that and he will end up buying a licence. If the poacher is under seventeen, it is not worth trying to prosecute, unless you are dealing with a persistent offender. If you are lucky enough to catch an adult, *always* prosecute – the word gets round and you will be left in peace for a while. For detailed information, the booklet *Law and the Angler* by Ron Millichamp, produced by the Institute of Fisheries Management, will tell you all that you reasonably need to know.

POLLUTION

I've left the worst predator of all until last – pollution. If this predator takes hold, there will be no need to bother about any of the other ones, prey and predators will all be dead and the poachers will have nothing to poach. Paradoxically, the more spectacular pollutions are the least of our worries, the fish come floating down the river belly up and everyone can see them. The water authority acts (although they themselves can sometimes be the polluter); the offender is identified, and pays substantial damages. It is regrettable, but some accidents are always going to happen, however hard people try to avoid them.

Fig 87   A lovely unpolluted river.

Much more worrying than these accidents is the insidious pollution which ultimately destroys all life. The most obvious one is the threat posed by acid rain. This isn't something that affects only the Scandinavian countries; it is destroying rivers and lakes in the British Isles as well. It seems that our industry, especially the coal-burning power stations, is the main culprit. These emissions can be controlled, at a price, but a price far less than the cost of liming and restoring all the polluted waters. According to one press report, it cost £1½ million to restore just one water, Loch Fleet in Galloway, and this is not a permanent solution. We are all entitled to clean air and clean water, but action will only be taken when the pressure of public opinion is great enough. The sooner more protests are made, by more people, the sooner action will be taken. Public opinion, and the decline in the profitability of farming, are already giving impetus to the improvement in farming practices and not a moment too soon. The River Waveney in Suffolk, for example, has lost much of its coarse fish population. The effluent from pig rearing units has been allowed to enter small drains and tributaries with the result that at certain times of the year ammonia levels are so high as to be toxic to fish. The problem is now being tackled, but it will take a long time to repair the damage and this kind of pollution is occurring in many parts of the country.

One feels a sense of hopelessness when writing about pollution, but the fight has to be fought. Wherever you suspect pollution, report it and keep complaining until something is done. Write to your MP and get your friends to do the same. I repeat – the pressure of public opinion is the only thing which will ultimately improve the situation, and at the moment we are a dirty, litter-laden country. Above all else, join the Anglers' Co-operative Association (*see* Useful Addresses), they will fight for us. Their most recent success has been the prosecution of the Thames Water Authority for polluting the River Thame. Anglers are by nature solitary creatures, but we must all unite against pollution and, by uniting, we shall win.

8
The Club

According to *The Concise Oxford Dictionary*, a club is 'an association of persons united by some common interest, meeting periodically for shared activity.' As far as I am concerned, the word 'shared' is the most important of all; the members must unite and share, if a club is to be successful. There is always a tendency, not just in fishing clubs, for a few members to play a very active part. The rest are much more passive, content to pay their subscription and to play tennis, bridge, football, or to go fishing whichever the case may be. This is quite normal. The problem is that, in a fishing club, the activity is rather more solitary. Very often members do not really get to know each other, and the passivity declines into apathy. At the end of the year, only a handful of members turns up at the Annual General Meeting, and the same committee is re-elected and runs the club for another year. Sometimes things work quite satisfactorily, but it is not a healthy state of affairs.

I believe very strongly that all members should be made to feel that they are part of the club, that this is *their* club. Even in a relatively solitary sport like fishing, there should be one or two functions a year, at least, where the members can meet socially and get to know each other. I have some responsibility for the running of two clubs. We manage about a fifty per cent turn-out at the AGM, not good, but a step in the right direction, especially as a large number of the non-attenders show their interest in the club by sending their apologies for absence; they are certainly not apathetic. If a good club spirit can be fostered and maintained, with the feeling that the club is something in which all the members share, then there is every chance that it will be a great success.

Where a few friends have come together to rent and improve a bit of fishing, very little is needed in the way of organisation. One of the group will usually, however, have the dubious pleasure of extracting money from the others, in order to pay the rent and other bills which may come in. There should be no trouble in agreeing among themselves how often they may fish or bring guests and what the catch limit will be. Working together to improve the water should be fun and the main requirement to make a success of the venture is that they get on well with each other. It follows fairly logically, therefore, that the bigger the club, the bigger the organisation needed. Some clubs, for example, actually own extensive fishing rights and their assets are very valuable indeed. One club, whose waters I fish quite often, owns fishing rights whose value is well over half a million pounds, and it is not salmon fishing, either. The club also rents a very considerable amount of water in addition to this. To run such a large organisation successfully requires men of a very high level of professional expertise. Though individual committee members take responsibility for the maintenance and day-to-day running of each individual fishery controlled by the club, a professional secretary is needed for the overall administration. The treasurer handles large sums of money and must look after the club's investments. This, too, is a job for a professional accountant. Finally, many expanding clubs will need the services of a solicitor and, possibly, a land agent from time to time. Lest you think that I am exaggerating the calibre of men required, I will give one recent example. A very large bank clearing job had to be undertaken. The secretary had to arrange for a couple of JCBs plus their drivers to be on site, together with an adequate number of chain saws and people competent to use them. The area to be cleared was divided into sections, with a committee man in charge of each section, the committee having been briefed in advance. A couple of hundred members turned up on the appointed day to help with the work; a field had even been rented to serve as a car park. The members were divided into teams, each team to work in its own section under the direction of the committee man. I leave you to imagine what a shambles such an operation would have been, had it not been professionally organised; just picture the typical club working party where every member thinks he knows best! As it was, the work was completed without a hitch, and much quicker than expected.

Most clubs will fall somewhere between these two extremes. If a new club is being formed, it will be run, at the outset, by those whose idea it was. A meeting will probably be arranged, in order to assess the amount of support which will be forthcoming and to ask for volunteers to help with the project, which is how I got myself into this mess in the first place! Right from the start, however, it is important to make members feel that they are very much part of the club, that they are all in it together. Sometimes there seems to be a gulf between committee

Fig 88   Even the secretary does some work, other than pen-pushing; I am seen here tidying up after the gale. The straight poles will be used for bank protection.

members and the rest, almost a 'them and us' feeling, and this is a poor state of affairs. The friendliest club of which I am a member doesn't bother about a committee at all. There are forty of us, or should be, but we usually end up with a couple more because the secretary/treasurer/chairman can't count! He does all the paperwork, having a most helpful secretary and all the resources of a busy office at his disposal. I look after the river, as I'm supposed to have infinite spare time because I am lucky enough to have early retirement. Whenever I need help, I phone up a few of the members and they are only too pleased to come along and lend a hand. At our AGM, which is more or less a social evening at the local pub, with a buffet paid for out of the season's surplus cash, members make suggestions for improvements and we do our best to implement them. They always insist that the two of us carry on with the job – they know when they are on to a good thing! The point is, we don't mind, as we know that our work is appreciated and that the members are anything but apathetic. I don't think that this sort of benevolent dictatorship would work in a larger and busier club, but we have no expansionist policies; our ambition is to make the very best fishery we can out of our stretch of river.

THE OFFICERS

I feel very strongly that, if a club is to be a successful venture, the considerations that I have been suggesting are most important. The first is that you are a club and that all members should feel that they are welcome and that all are in it together. The second consideration is that the club should not be *over* organised; an excess of officers and committee members often leads to inefficiency and slow decision making. The management team should be as small as necessary, rather than as large as necessary, to run the club. The final consideration is that the officers, especially secretary, treasurer and fishery manager should be fully capable of doing the job effectively – not an easy matter to decide, but people who are successful in their work will probably do a good job for the club.

The Chairman

Before discussing these three officers, let us consider the chairman, or, if you live in a non-sexist area, the chairperson. At some club meetings, I'm not sure whether a chairman or a referee is required; it is surprising how much heat such a placid sport can generate. Certainly the chairman must be someone both well known to all the club members and well respected by them. He need not play a very active part in running the club, though he can be a useful peacemaker at committee meetings. It is at the AGM, especially if the club is a fairly large one, that the chairman's role is essential. The club secretary will have prepared the agenda and done his best to ensure that everything runs smoothly but, on the night, it is the chairman who is in charge and, apart from their reports, the less one hears from the rest of the officers, the better. This is the time when all the members can make their comments, criticisms and suggestions, all of which should be addressed through the Chair. I am sure that we have all been to meetings where the chairman is not in control, and the evening degenerates into an argument between two or three vociferous members,

Fig 89   *Frank, our ex-chairman, patches one of the boats.*

addressing each other and ignoring the Chair completely. So I repeat, the chairman must be known and respected, well able to control the meeting and to give all members a fair chance to speak; not an easy position to fill.

The Secretary

The administration of the club falls to the secretary and the treasurer; in a small club this may well be one and the same person, but I will deal with them separately. Provided that the club is not too large, the secretary needs no great talent, which is just as well for me, as I'm secretary of one club; reliability is the virtue most needed. Letters must be written and answered promptly, the membership list must be correct and up to date, the agenda for meetings has to be prepared and minutes kept and written up – a pretty dull job! In a large or expanding club, where new waters are being sought and negotiated for, I feel sure that the secretary should be a professional man, with secretarial resources at his disposal and that he should be paid for his services – a far cry from the traditional 'Hon. Sec.' like me. There is one aspect of a secretary's work which tends to be neglected and that is the job of keeping the membership informed of the club's activities, and keeping them interested in them. It is not enough merely to issue a membership card and to give due notice of the next AGM. Little wonder that members become apathetic when this state of affairs is the norm. It costs a little in time and money, but the occasional club letter is a great help in fostering a good club spirit. I think that it is best if this letter is written by the secretary. I tend to send out five or six letters a year, but even if there is only a midseason report on the fishing, it is a useful exercise.

The Treasurer

As for the treasurer, what a rotten job! I have done my stint in my younger days, at both tennis and badminton clubs, and I hate handling other people's money. On one occasion, when the books failed to balance as usual, a strange item appeared in the income and expenditure account as 'unexplained surplus £4.50'. Even then, the club didn't take the hint and find a new treasurer; as long as you make a profit, all is well. Reliability is the only virtue needed in a club of moderate size, as it's merely a question of collecting the subscriptions, paying the bills, and producing a satisfactory income and expenditure account at the end of the year. In a large club, however, whose income might run into tens of thousands of pounds, the treasurer should be a qualified accountant and auditors must be appointed to check the books and the balance sheet. It is not unknown for a club treasurer to misappropriate the funds and end up in prison, even in a fishing club, and we like to think that we are an honest lot, except for a slight tendency to exaggerate. In the large club, a professional accountant is vital; apart from the skill of keeping accounts in the correct professional way and producing a proper balance sheet, there will almost certainly be club investments to control as well.

The Committee

The club will probably have a committee, but in my view, the smaller the better, as there will be fewer people to tell each other about 'the one that got away'. I plead guilty of this, and, in one of our clubs, the chairman is as bad as the rest of us. Unfortunately, committees tend to be elected and there is always the risk that the more popular or, at worst, the more vociferous candidates may be chosen. If I had my way, the committee members would be volunteers, who have something valid to contribute to the work of the club. Running a fishing club calls for a lot of practical skills and builders, plumbers, electricians and the like make ideal committee members. People in business, too, can be a tower of strength, especially in dealing with the paperwork. Wonderful machinery is available in offices nowadays and all the typing, duplicating, and photocopying can be taken care of, keeping the membership well informed. A club should not expect to have these benefits given free of charge, but will still make a great saving in cost. I think it helps if committee members do about a three year stint and then stand down for a while, so that new blood and new ideas keep coming in. Continuity in management is needed, but the whole thing should not be allowed to fossilise. In the club of which I am secretary, there is usually about one committeee change every two or three years.

President and Vice-President

As for the other officers, presidents and vice presidents may be classed as 'optional extras', but it is a good thing for a club to be able to bestow some sort of recognition, be it to a landowner or to some other person who has helped the club. People should know they are appreciated and are not taken for granted, though I did just that to a club member only recently. He always takes the minutes of the AGM for me; I had made a note to thank him for doing so, but had forgotten to ask him if he would!

Fishery Manager

I have deliberately left the fishery manager until the end, but his role is vital if the best use is to be made of the water available, especially the type of water discussed in this book where a big effort is being made to establish a good trout fishery on what, in farming terms, would be called 'marginal land'. Secretaries and treasurers can be pressed into service, plied with alcohol before the AGM and find themselves in office before they have realised their mistake! Not so the fishery manager. He must be the enthusiast who really *wants* to do the job. The problem is that he

Fig 90   Jack Fitt, the Fishery Manager and the driving force of the club.

may not have any experience; I didn't – another reason for this book.

On still waters, in lowland areas, the problems facing the fishery manager, whether he is a professional or one of the club's officers, are very similar – trout fishing has to be provided on what is not naturally a trout fishery. On the river, the club fishery manager is probably confronted by a different set of problems from those encountered by the professional river keeper, who is almost certain to be looking after water which is naturally quite good trout fishing. The club fishery manager may well be starting from scratch and it is his drive, enthusiasm and ideas which are essential. You cannot really do without a madman like this, if you are to develop your waters into highly successful fisheries. The rest of the officers will act as a kind of brake, so that instant insolvency can be avoided and the tradesmen on the committee will translate the more lunatic schemes into practical reality. I may seem to exaggerate, it's a fisherman's privilege, but a leader is vital. On the largest of our lake fisheries, the fisheries manager is such a man, backed up by a committee of competent tradesmen. As a result, we enjoy superb trout fishing at modest cost; the proof of the pudding is the length of the waiting list and the scarcity of resignations.

THE RULES

Now that there is some sort of administration, I suppose the club will have to have rules but, as with committee members, the fewer the better. Members' tastes, interests and styles of fishing differ enormously, especially in a large club; just because one member prefers upstream dry fly fishing, there is no valid reason why that preference should be made the rule for everybody. For some lake fishermen, lures and sunk lines are anathema, but plenty of others are quite happy to spend the whole day dredging diligently. Let me confess now, I'm quite content to drown worms from time to time, if the rules allow it, but I'm not very keen on spinning; however, that's no reason why spinning should be banned. In general, fly fishing only is the rule on most trout fisheries, possibly because it puts slightly less pressure on the stock of fish. Even so, there is no reason why fly fishing should be some sort of 'sacred cow'. In one club whose waters I fish and which controls quite a long stretch of river, both spinning and bait fishing are allowed in some sections. In fisheries where several lakes are controlled, one of them is sometimes kept as an 'all methods' fishery and, on other bigger waters, all methods of fishing

Fig 91   The landlord plays his part on this fishery – no DIY job, this one!

are sometimes allowed from one of the banks. Even on 'fly only' fisheries I'm surprised how intolerant some anglers are of others' methods; I don't like stripping in a lure at a great rate of knots, a matter of personal temperament, not morality, but I enjoy rowing and will happily trail my flies behind the boat during quiet spells though, alas, that is forbidden on most reservoirs; I don't really see the difference between the two. Changes are coming, especially in some of the large reservoirs which hold big stocks of good quality coarse fish. The new English record pike has already come from one of these waters, namely Ardleigh Reservoir. Towards the end of the trout season, all methods are allowed on Ardleigh, even for trout, and then the coarse fishing continues on through the autumn and winter. The fishing is good enough to justify the fairly high price of the day ticket and the management has discovered that the bait fishermen generate very useful extra income, without spoiling the trout fishing, which, in any case, depends mainly on fish stocked throughout the season. This extra income may not only make the difference between profit and loss, but can provide much-needed cash to enhance the trout stocks. So, before you frame your rules, a plea for toleration. Don't let personal prejudice cloud your judgment, and remember that tastes differ.

In framing the rules, three basic matters have to be decided; the number of members, the bag limit and the fishing methods.

Members, Guests and Day Tickets

The more members there are in a club, the cheaper the fishing will be. All of us, alas, have some degree of selfishness and it is annoying to find one's favourite stretch already occupied. Even so, I mentioned the word 'sharing' at the start of this chapter. The fisheries with which I am associated are not commercial ventures and the aim is to provide moderately priced trout fishing for as many people as possible. A lake can accommodate far more anglers than a river, without their getting in each other's way. In order to lessen this problem, the river fishermen are limited to a certain number of visits per week or per month, depending on which fishery they belong to. Even so, we avoid giving members a fixed day or days on which they may fish. This does risk the occasional busy day, but we have found that in practice this rarely happens and the free and easy approach is worth much more than the occasional inconvenience. On the lakes, the members can fish as often as they like, subject to a catch limit for the week. Apart from the opening day stampede, which is more of a social occasion than a serious fishing expedition, we have never found that we are overcrowded. In order to give a definite idea of numbers, the largest of the lakes is 25 acres in extent and also has a big island which provides extra bank fishing space. There are 110 members, fishing as often as they like, and no problems are encountered. The busiest of the river fisheries has 40 members, allowed to fish four days a month – not one day per week. There are about two kilometres of fishing available, but the arrangements seem perfectly satisfactory.

Some arrangement has to be made for guests; most of us actually have friends and like to be able to offer them some fishing when they come to visit. In the Salmon and Trout Association, one or two tickets per day are available to Association members visiting the area, or to local members who are not members of the club itself. On the club fisheries, the arrangements vary. In one of the biggest clubs I know, members are allowed two cheap tickets a month, except for April. They may also

Fig 92   No guest here, but they often do get roped in to help.

bring a friend and share a rod, one day a week, if they wish. To avoid the risk of abuse, no guest is allowed more than six visits a year to this fishery and must always be accompanied by the member who sponsors him. This stops rod sharing, in the sense of paying for one membership but having two members fishing. Another club I know, in a different area, issues its members with half a dozen free tickets as part of the privilege of membership but, again, the guests must be accompanied by the members sponsoring them. Only one of the clubs I know of allows unaccompanied guests but these guest tickets are very limited in number and will only be given to people whose conduct is utterly trustworthy.

Day tickets, of course, are a separate subject, and there is a big distinction to be made. In holiday areas, where the fishing is used to attract the tourists, it is essential that suitable arrangements can be made, though tickets may have to be limited to prevent overcrowding. On the type of club waters with which this book is concerned, however, I am totally opposed to the sale of day tickets. There is the risk of attracting the 'fish hog', who is after the fish and nothing else, and over whose conduct the club has little control. The sale of day tickets does nothing to enhance a good club spirit. The case of the Salmon and Trout Association fisheries in this area is slightly different. These were set up with the direct intention of encouraging people to join the Association, and the day tickets are only available to members so, even if they are not members of the club which looks after the fishery, the day ticket holders have a real interest in the well-being of the water.

There is no guarantee that membership and guest arrangements will be fully satisfactory right from the start, but rules can easily be adjusted. Too many members, and the surplus can easily be shed by natural wastage until a more suitable number is reached. If there are too few members, there should be no problem in recruiting more, provided that the fishing is good.

Bag Limit

Having settled the number of members and guests, then it is a question of fixing the bag limit. Catch and release is discussed more fully elsewhere. On some waters in Europe and America, it is the rule, not the exception, just as it is in coarse fishing. If the rule that all fish caught must be killed is adopted, as it is on the majority of trout fisheries in England, then the limit has to be set high enough to make a visit worthwhile. I certainly would not want a round trip of fifty miles or so, if I knew that a brace of fish was the limit and that I might possibly catch that limit in the first two casts. Our lake fisheries do not permit fish to be returned and the limits vary between four and eight fish per week, thus allowing the angler who lives close at hand the possibility of making several visits and yet leaving a good day's fishing for the angler who has much further to travel. Rules on the rivers vary; one of my clubs has a limit of a brace but any number of fish may be returned and the use of barbless hooks is encouraged. The smallest club of all allows its members thirty fish a season and relies totally on the honesty of its members not to kill more, a free and easy approach which I am sure is not abused. Once again, the bigger the club, the more likely it is that there will be problems. A few wretched individuals are just fish hogs and will not abide by the rules. The only answer is to expel cheats whenever they are caught. I know one club secretary, who is also paid as a bailiff, who quite often watches the water through his binoculars. In this way he has managed to weed out some undesirables. I think that much stricter rules and enforcement are needed in large clubs than in the small ones, where

Fig 93   Graham, one of our members, hard at work.

all the members know each other. I will talk more about bailiffing and enforcement later.

Fishing Method

The last of the three main rules to consider is fishing method. As I have already suggested, much depends on the type of water available and its extent. Despite my fondness for a bit of worming and my plea for toleration, fly only is the most sensible rule for most trout fisheries. Some rivers insist on the rule of upstream fishing only. I practice this as a matter of personal preference, but I can see no valid objection to fishing across and down, though I would limit the size of fly to be used.

The greatest care has to be taken when framing the rules on fishing methods, so that honest anglers are not encouraged to become cheats. I'll give one example. A very nice sea trout fishery in the Borders allowed the use of the upstream dry fly, but the nymph was taboo. The thinking, if any real thought there was, had been that the upstream, sunk fly would be used to foul hook fish. Commonsense would merely have banned the use of double or treble hooks and limited the size of hook permitted. I used my size 12 and 14 weighted nymphs to quite good effect, but I was, in point of fact, cheating, though it would be hard to imagine tackle less likely to foul hook fish.

The use of boats may also be considered under fishing methods. The fisheries that we are dealing with are not generally large enough to permit the traditional style of fishing the drift, which would almost certainly inconvenience the bank anglers. On a fairly large lake, where much of the water is out of range of the bank, it is not a bad idea to provide a certain number of buoys to which boats may be moored, thus making full use of the available water. Rules need to be framed to allow members a fair chance of booking a boat, should they wish to go afloat, and of limiting the time they may spend out in the boat.

Minor Rules

That leaves only the minor, local issues to be settled – booking in and out, catch returns, parking, access, radios, dogs. I certainly don't go fishing to listen to the radio, mine or anybody else's, so I would ban them on the grounds of inconvenience to others. Dogs I would also ban, though most unwillingly. Some members are the owners of badly behaved brutes and, because of them, all must suffer. This is a pity because some dogs are not only well behaved but also fanatically keen on fishing. We met one such enthusiastic hound recently in Scotland. He marked every fish that jumped, watching the river intently all the time, and whenever his master hooked a finnock, he would go and stand by his side, quivering with excitement, until it was landed. I have also read of dogs which landed their master's fish, so that a landing net was superfluous except, occasionally, for landing the dog! Such paragons are the exception, so 'no dogs' is the safer rule.

Club members should be encouraged to do their own bailiffing and rule enforcement, whether a keeper is employed or not. The keeper cannot possibly be present all the time. If a good club spirit has been fostered, the only problems will be caused by trespassers and the occasional 'black sheep'. In a good club, the members will abide by the rules and take a pride in their waters. I was fishing as a guest on a river in Scotland, a club water, and my hosts kept a beady eye on me to make sure that, having waded out from their own bank, I didn't happen (in my enthusiasm, of course) to stray a few yards downstream into enemy territory, where the grass is always greener!

SUBSCRIPTION

All this has, of course, to be paid for and none of the clubs in this area is in the business of providing exclusive fishing for a limited number of wealthy members. Working out the annual subscription is a simple matter, provided that a suitably pessimistic view is taken. Add up all the known costs, then assume that most of the things that can go wrong will, so add at least another ten per cent to the grand total. Divide this horrendous sum by the number of members and that is the subscription, except for a bit more pessimism; round it up to the next £10, so that for instance, £124 would equal £130. Crude mathematics, but it is better to have more rather than less! You shouldn't need me to go through all the likely expenses, but I will quote the items listed on one club's expenditure sheet: stocking, rearing, food, leases, wages, taxation and insurance, secretarial services, amenities, affiliation fees, development and maintenance. That lot is quite normal; some years there may be legal fees as well if a new agreement has to be negotiated or a new water acquired. It is wise to aim to build up a reserve fund in the course of time, so that real emergencies (the loss of one's fish through disease, for example) can be coped with more easily. The opportunity to buy fishing rights should also be born in mind. The club subscription is really no problem at all, provided that members think they are getting value for money. Some of ours even think they ought to pay more!

WORKING PARTIES

Working parties play a big part in running a club successfully and in keeping the subscription as low as possible. I might even have included them in the list of club rules as, in some clubs, attendance at working parties is a condition of membership. I belong to one such club and, during the last few years, I have worked more than I have fished. This is no hardship, as the annual subscription is small and the working parties are enjoyable. Another much more expensive club, whose waters I often fish as a guest, also has this rule; there is much work to be done, if the fisheries are to be maintained to a high standard. Such a rule has to be enforced fairly, so that each member gives according to his ability. A member makes a contribution just by turning up, even if he is not fit enough to do a great deal. One of our lady members is a most reliable scorer on stocking days. She keeps a meticulous tally of the number of browns and rainbows, so that they almost have to ask her permission before going into the lake! Where working parties are voluntary, there is the perennial problem of members who always seem to find time to fish regularly but never have enough time to lend a hand to improve their fishing. I favour compulsion, a typical old-fashioned schoolmaster's view!

Despite this last comment, I believe in as free and easy an approach as is compatible with the size of the club. Rules should be kept few and simple, and the subscription as low as is realistically possible. In this, I seem to be at direct variance with Alex Behrendt, whose view is that such a system, where members can come and go as they please, is most unlikely to be successful, and he would anticipate a rapid turnover of members, poor finances and indifferent fishing. My experience has been exactly the opposite. On our lake fishery, with 110 members, we have very few resignations indeed, two or three a year. Though there are plenty of competing fisheries, both club and commercial ones, the waiting list, which I update every three years, is enormous; it will be the next century before some have a chance to join! I am sure that, if a good standard of fishing is provided and that you have a genuine club, where

*Fig 94   A working party at the concrete pen – 'long johns'
are obligatory.*

Fig 95 We have no trouble with working parties; this little dumper has been invaluable to us – kindly provided by the landlord of this fishery, Cyril Rogers.

the members are prepared to take the rough with the smooth, you will be successful. One year, we had a pre-season disaster. We had just stocked the lake when there were enormous floods and we lost almost all our fish. One member asked for his money back, so we refunded most of his subscription. He has no chance of getting back into the club due to the long waiting list, and has been kicking himself ever since!

9

The Proof of the Pudding – Two Case Studies

ABBOT'S HALL

Abbot's Hall Fishery is quite an attractive stretch of lowland river, approximately two kilometres (1.2 miles) in length, which is rather shorter than I would really like. It is about twelve metres (44 feet) wide on average and I can walk down about half its length in chest waders. Shepherd's Hole, named after a previous owner, is about two metres (7 feet) deep, but this is easily the deepest spot on the whole stretch. The fishery was set up in 1981 and is the brainchild of David Clarke, who owns one bank and has leased the other from the National Trust. Prior to this, the water authority had rented the river, which was very good dace and roach water until coarse fish stocks in the area declined sharply during the 1970s. The authority had been tipping in a few trout every year and selling day tickets, but had not really done any work to improve the habitat. This was certainly not a very profitable venture and in 1980, as the financial climate was becoming harsher, the authority gave up all its leases. When this happened, Mr Clarke had the far-sighted idea of offering the fishery to the local branch of the Salmon and Trout Association as a means of recruiting more members, attracted by modestly priced trout fishing.

The local branch of the Salmon and Trout Association had been reformed in about 1978, but there were only a couple of dozen members. The offer was accepted, a doubly generous one in that, in order to help the club get started, Mr Clarke was to pay the National Trust's rent for the first year, and he still gives his water rent free, on condition that we work to improve it. He can scarcely have imagined the success of his initiative. Our branch of the Salmon and Trout Association now runs five separate fisheries and is one of the largest in the country, with well over 800 members. Angling is essentially a fairly solitary sport, but as I have said, it is vital that anglers, coarse and game fishermen alike, give their support to the bodies which exist to protect our common interest.

But enough of angling politics! A meeting was called in March 1981, so that those who might be interested in forming a club could get together and have a look at the river. That gave us all of two or three weeks to get something organised before the start of the season! I've no recollection of how I heard of the meeting or of how I came to be there. All I know is that, as usual, I couldn't keep my big mouth shut and ended up as one of the new recruits to the Association and with a river to look after into the bargain. One member acted as chairman/secretary/treasurer, and still does. He extorted money from us and went off to order some trout, more in hope than in expectation. Another went off and designed and built a splendid bridge, so that we could have a river crossing at the downstream end of the fishery (there was already a

Fig 96   Goff's Bridge.

wooden bridge near the upstream boundary). I was left to work out a plan of campaign to improve the river and it also fell to my lot to organise the 'slave labour', yet another instance of the blind leading the blind.

The first step was to survey the river and its fish stocks. The water authority was most co-operative and organised an electro-fishing operation. The results were very surprising. The river teemed with fish or, to be more exact, eels. As for proper fish with fins, tails and scales on them, no such luck. In the couple of kilometres of water that we fished, there were no roach at all, one rudd, one perch, a handful of dace and a few trout, the survivors of earlier stockings. There were also a couple of dozen pike, doubtless a sub-species with a special fondness for eels, lampreys, and moorhen chicks! Despite this almost total absence of fish, apart from the aforementioned eels, the water quality was excellent, 1A, so there was no apparent reason why trout should not do well, even though the water authority's efforts had not been spectacularly successful.

The survey of the river bed was easy enough – that is how my luckless wife found herself sitting in the back of the punt, with pencil and paper and a rough sketch map of the river. We drifted slowly downstream, while I prodded away at the river bed with a long angle iron, which had been gaily painted in feet and inches to ensure the accuracy of our highly scientific recordings. The results were predictable to the point of monotony – mud, anything up to a metre (3 feet) deep, covered the bed of the river, and the angle iron came up black, stinking and horrible. There was some good news, however, as underneath the mud the river bed was mostly gravel, gravel which was only exposed on a couple of quite fast-flowing shallows. My wife dutifully recorded the depth of the water and of the mud, which at least gave me some idea of where the biggest improvements might be made most quickly.

The trout which had been ordered were duly put into the river and, as you might expect, provided only limited sport, especially as the first stocking had been made much earlier in the season than we would ever stock now. By mid-summer, the river was absolutely choked with weed whilst the banks, on the other hand, were almost devoid of cover. My plan of campaign was obvious enough; clear the mud, reduce the weed growth but, at the same time, introduce better quality weeds and, finally, increase the amount of bankside cover. During the first two seasons, the catch rate remained rather poor and the high spot of the first year was the appearance of a large golden orfe, which had escaped from somebody's pond in a flood. It was at least a two pounder and it flaunted its brassy bulk up and down the river until it disappeared during the summer holidays, probably falling victim to some poacher's worm.

We were not daunted by our lack of immediate success – the neglect of years cannot be put right in five minutes. We toiled away during the first two years, doing all the things that I have suggested in chapter 2. Obviously, some of the jobs were not done as skilfully as we would do them now. This book is, to some extent, the fruit of our mistakes and I hope that it may help you to do a good job, first time around. Plenty of club members joined in with the work and I had a great ally and helper in Mr Clarke's gamekeeper. He has the advantage of being strong in the arm, without being weak in the head. He often showed me the best way to translate my ideas into practical reality. In fact, Robert, the keeper, is such an ally that I heard the owner complaining, loudly enough to make sure I heard, that his keeper spent more time looking after trout than pheasants and that even the farm foreman was busier making things for me, than working on the farm!

Fig 97   The upper stretch.

Not only was success slow to come, we also had to endure complaints from people living downstream, who seemed convinced that the river was being blocked off and that the town must inevitably be flooded as a result of our labours. The water authority engineers paid us a visit, in response to the complaints, but, apart from suggesting a couple of modifications, they were favourably enough impressed to help with some of the improvement work and to survey the river for us. The survey underlined just how lowland our river is, with a fall of much less than a metre (3 feet) in over two kilometres (1.2 miles) of river. I value the help of the authority and there is much more co-operation between the fishing and engineering branches now than seemed to exist in the past. Modern university courses may cover the field better, but many engineers have not yet received adequate training in the ecological and conservation impact of their work. Many are unaware, for example, of the simple devices that can be used to improve the habitat without harming drainage, which must be their prime concern. In my modest way, I feel that a visit to Abbot's Hall would provide them with a useful bit of training!

Though the apparent results during these early years may not have been very good, the river was working for us all the time; more and more clean gravel was appearing as the current carried the mud away, in fact, the stretch downstream of us is due for dredging this year, though I don't expect that is entirely our fault! A lot of unwanted weed had been cleared out and the faster current ensured that, in many places, it didn't come back again. Even the trees that we had planted and which escaped the onslaught of the animals and, even more lethal, my strimmer, were big enough to be seen and would soon be providing a little shade. War was waged against pike and eels, the eels providing a small financial bonus for the keeper, who helped the eel trapper. The contents of the nets were always interesting. One day there were a few tench and a couple of bream, species which we had never thought were in our stretch at all. Crayfish are quite plentiful and the nets always come up absolutely covered with caddis, in their little twiggy houses, so there is a good supply of trout food.

Then in the third year, all our work started to bear fruit; Fig 99 shows clearly how the fishery has progressed. As can be seen, members and their guests now pay far more visits than they did in the early days. We also stock with far more fish than we did at the outset, but while we have only doubled our stocking levels, the total fish caught has gone up fivefold. The fish settle into our stretch of river very well and plenty can

Fig 98   The moment of triumph.

Season	Number of members	Number of rod visits	Trout killed	Trout returned	Total caught	Number stocked	Average per rod visit	Fish over 2lb
1981	40	101	86	26	112	400	1.1	0
1982	40	110	48	69	117	450	1.1	0
1983	41	164	106	125	231	550	1.4	0
1984	41	235	177	161	338	625	1.4	9
1985	40	262	249	300	549	650	2.1	11
1986	41	294	311	214	525	650	1.8	17
1987	42	344	363	255	618	750	1.8	30

Fig 99   Abbot's Hall Fishery – an overview. Notes: a)
Brown trout only are stocked, three times a year, in mid-
May, late June and late July, except for 1981, when there
were only two stockings. b) For the first three seasons, only
trout of about 12in were stocked, but since then there has
been a mixture of 12in fish, larger ones averaging over
1½lb, and a few big fish; the largest caught was 4¼lb.

be seen rising after the season has ended. We keep the catch limit low, as we wish to encourage the sport and not the fishmongers. We have certainly succeeded in this, as many members do not even kill their brace, being quite happy to catch and release.

One interesting feature is that our river is a very late one. Weed growth is so slow that little is in evidence until well into June. This is the reason why our initial stocking is so late and, in fact, the water authority has altered the dates for the brown trout season, which now extends to the end of October. It is difficult to assess the value of this change at the moment, as 1987 saw the river in almost continuous flood from late August onwards. Certainly there are excellent fly hatches in the autumn months and I have seen fish rising well, even quite late on in November. The monthly totals for one season (*see* Fig 100) illustrate this pattern quite well.

The most disappointing aspect is how few stock fish overwinter, despite all the work which has been done to improve the habitat. Many of the survivors can be seen spawning, but then they seem to disappear for ever. My only comfort is that this pattern seems to be repeated with stock fish on many other river fisheries, so it's not just my problem. Even so, I have a stubborn streak, and next season we will be putting in a few smaller brownies, which someone is rearing for us, a little higher up the valley. These fish, coming from the same water, may stay with us longer.

1985 season	Rod visits	Trout killed	Trout returned	Average per rod visit
April	9	1	0	0.1
May	23	16	11	1.2
June	45	22	29	1.1
July	54	63	89	2.8
August	63	84	141	3.6
September	68	63	30	1.4

Fig 100 The 1985 season at Abbot's Hall.

Abbot's Hall Fishery has been a most successful venture, more than fulfilling all our original aims to increase branch membership by providing modestly priced trout fishing. We now have quite a long list of Association members, who would like to join the club. One day ticket is kept available every day, so that everyone will have a chance of fishing the water if they want to. In 1987, for example, over thirty guest tickets were sold. Our success has encouraged the branch to form two more clubs on other rivers in Norfolk, both of which are prospering, and we have two lake fisheries as well. The real point, however, is that I can think of three more small rivers within twenty miles of here where I am certain that I could create a decent trout fishery. This type of opportunity must be available in many other lowland regions, given the goodwill of the landowners and a few idiots like me to help organise the work.

Problems will always occur, and our outlook at Abbot's Hall is a bit troubled at the moment. A little tributary has been dredged too deeply and iron salts are leaching out and clouding the main river. Fortunately, we are only slightly affected, but there has been a marked decrease in the growth of ranunculus and we may also be suffering from a slight increase in the acidity of the water. This pollution is not toxic to the fish, but it is bad enough to be spoiling the fishing for the club who rent the water upstream of us, where the little stream enters the river. The water quality people are investigating, though without much enthusiasm, as no fish are being killed. Their view seems to be that the nuisance will diminish gradually and that we must grin and bear it. Clearly, nature will eventually take its course, but that could well be later rather than sooner, and we would rather see some positive action taken.

The important thing, however, is not to be deterred. Setbacks and mistakes are inevitable, but the good work must go on. This year, a

Fig 101   A pretty corner of Abbot's Hall.

couple of new groynes are planned and a couple more need renovating, jobs which ought to make a couple of hundred metres of water more productive. A tonne of Siltex is waiting in the barn, until the high flow eases a bit. The extra chalk may help to maintain the pH, which is threatened by the iron salts, and it will certainly help in the battle against mud. The work is never ending, but I'm told that the conservation officer from a neighbouring water authority is seeking to increase his practical experience and is coming to stay for a week. Poor chap, he doesn't know what he is letting himself in for!

LYNG

Stillwater trout fisheries are two-a-penny these days; everybody seems to be stocking any puddle with trout, but this most certainly wasn't the case back in 1969, when the fishery at Lyng was founded. Grafham Water had not been open long and it was the impact of the superb fishing there that led to the formation of the club, as there was now a local demand for trout fishing. For many of us, Grafham was nearly three hours' drive away, no trouble at all on the outward journey, but often a nightmare coming home after a long day's fishing. I certainly had to pull off into a layby and go to sleep on more than one occasion – and some people still think that fishermen sit down all day doing nothing! So, back in 1969, the founding of a small stillwater fishery was quite an adventure. There was nothing like the store of experience which is available now.

Lyng is quite a big gravel pit. There are rather more than twenty-five acres of water and the amount of bank space available for fishing is greatly increased by a large island. There are other small islands providing a little extra fishing space, but these are of more interest to the ducks and geese than the fishermen. The water is fairly shallow, two to three metres (7 to 10 feet) deep, though there are a few spots where depths of four and five metres (13 and 16 feet) can be found. A small river flows close to the west and north banks of the lake and a stream runs along the whole of the southern boundary, very close to the lake itself. Two other gravel pits on the site, one of them very small, are kept as coarse fisheries. The water quality, the pH and the food supply are all good, without being in any way exceptional; the only exceptional thing about Lyng is the fishing!

This is not one of the Salmon and Trout Association fisheries, having been founded long before the local branch expanded so much. Even so, we are club members of the Association and of the Anglers' Co-operative Association, as any responsible club should be. The membership is limited to 110, in order to give the fishermen adequate space, though opening day is a bit of a crush. It is in no way an exclusive club, as the membership fees are kept as low as possible – the whole season at Lyng costs less than two days' trout fishing on the Test, for instance. The membership represents a full cross-section of the community, which is possibly one of the reasons why the club is so successful. There is always somebody who knows how to do a job or how to get something that is needed. We even have a professional sea fisherman, so our netting team has had excellent instruction!

When the fishery opened, the angler was confronted with a very bare-looking expanse of water, as gravel extraction had only just been completed and there was very little in the way of trees and shrubs. The banks themselves were rather rough, as they had not been properly graded and they were still settling and drying out. The fishing followed what seems to be quite a common pattern for still waters, good fishing to begin with, then a period of decline. One reason for this decline was undoubtedly the coarse fish explosion. As I write this, I can look at a couple of two pound roach that my friend set up for me. For two or three years, the autumn roach fishing was incredible and, when conditions were favourable, super roach could be caught steadily all day. Pike were so numerous that one afternoon my son and I caught a couple of dozen small ones in under two hours. When the fish spawned in May and June the margins were alive with big roach and huge bream.

As can be imagined, there was a steady turnover of members during this period, but those of us who liked coarse fishing weren't complaining, especially as the roach fishing in the river was also excellent. The minutes of the AGM and committee meetings during that period give fair space to fish trapping and netting operations. These met with mixed success; apart from one huge haul of bream in the fish trap, I can't really believe that we made serious inroads into the coarse fish population. Other factors had a big influence; the perch, for example, disappeared during the period when a perch disease swept the whole country. There was a very large build up in the wildfowl population, especially the ducks, as the lake is used for shooting during the winter. I don't know whether ducks eat spawn, but I'm always amazed at their appetite for small fish, though these are usually the stunned and half-dead ones left about after a netting operation. The constant introductions of trout may also have played a part in the decline of the coarse fish, because trout are now very

Fig 102   Evidence of success.

much the dominant species. Whatever the reason, the coarse fish population is now drastically reduced and we no longer see vast numbers of fish during the spawning season.

The season's totals (*see* Fig 103) since the fishery began show how the lake went through a rather lean spell, a spell which we hope is now firmly behind us. Disasters will inevitably occur again, just as they have done in the past. Some of the poor years shown are quite easy to explain. 1971 was the year of the eye fluke; 1979 saw the loss of almost all our stock fish in disastrous floods, and in 1980 large numbers of stock fish were again lost.

A set of figures like this gives a fair picture of the success of a fishery, or the lack of it, but it most certainly doesn't tell the whole story. The history of Lyng is divided roughly into three parts: the early years when all the fish stocked came directly from the fish farm; the middle years when we were struggling with the problems of rearing our own fish and when, incidentally, a new owner bought the fishery; and finally, the last seven years, during which we have learned from our mistakes and profited as a result.

Season	Total stocked	Total caught
1969	5000	3054
1970	5000	3067
1971	5000	2890
1972	5000	4181
1973	5265	4342
1974	5000	3178
1975	5000	3346
1976	4700	1914
1977	5000	2987
1978	5630	2840
1979	4915	2373
1980	5500	1693
1981	5050	4110
1982	5076	4520
1983	5395	4675
1984	7000	4999
1985	5566	4871
1986	6824	6080
1987	6275	5961

Fig 103 Lyng Fishery's yearly totals.

The late Ken Smith played a big part in setting up the fishery and was its first manager. Ken was a superb coarse fisherman and had actually won the All England Championship but, like many coarse fishermen, he enjoyed trout fishing as well. Under his guidance, the club flourished and there was only one lean season, when all the trout had eye fluke and, from late July, we used the lake as a coarse fishery. One day I was fishing near Ken and had been broken about three times by what I thought were big tench. The fish had plunged deep into the blanket weed which covered the bottom and that was that. The same thing had happened to Ken but he solved the problem in style and landed an eel of well over four pounds, one of the biggest I have ever seen alive. During these early years, all the trout came directly from the fish farm and, though a few larger ones were stocked, the tendency was for most of the fish to be very similar in size. Despite this one quibble, the fishing was most enjoyable. There were good fly hatches; in fact, some June evenings you were covered almost from head to foot with caenis, and the trout could be seen everywhere, sipping them in and steadfastly ignoring everything else.

This ended the first phase, the more so as the fishery changed hands the same year, and was bought by Mr Cyril Rogers, head of a large building firm. He was ambitious, and still is, to develop the full potential of the site. Since he bought it, the whole place has been transformed. The first thing he did was to level the banks and have them properly grassed, a job which made an enormous difference to the appearance of the fishery, and made its circumnavigation much less hazardous. Our old bridge to the main island was effective, though slightly hair-raising. It floated on a line of old oil drums, so that you bobbed and wobbled your way across. This was a journey not to be recommended to those who suffered from seasickness or were of a nervous disposition, especially when one or two of the drums started to leak and the whole affair took on a pronounced list. This was replaced by a fine iron bridge, strong enough to take machinery across to the island if need be, and certainly the kind of job which would have been well beyond our means. Cyril helped with other major work during the next few years, including the construction of pens, to help with the trout rearing. A few years later, the site was given an award by the Sand and Gravel Association to mark the quality of the reclamation work.

This second phase is very much the story of two people and a band of very hard-working club members. Cyril Rogers had a big impact on the landscape and the major construction work. Jack Fitt, the club treasurer,

Fig 104 The advantages of rearing your own trout.

now became Fishery Manager though, in our haphazard way, it was several years before he was actually given the title. I cannot praise him highly enough. Any club needs a Jack Fitt, if it is to make the most of its fishery! Jack realised that, if the fishing was to develop to its full potential, it was essential to rear our own fish, or at least the bulk of them. His arguments were that there would be a big cost saving, plus the possibility of stocking with trout much more varied in size. We would also be more independent of the suppliers, some of whom were having troubles of their own and were finding it difficult to guarantee delivery. With our own fish, the stocking programme could be much more flexible. Finally, fish reared in our own water should have a much better chance of thriving when liberated than trout which had come from a very different, spring fed environment.

Life was not easy for Jack during this period, as he encountered considerable opposition, many members being totally against the idea of growing on our own fish. As fish rearing progressed, there were a series of disasters and everything seemed to conspire against him. The inevitable 'I told you so' was heard far too often. Jack's enthusiasm and stubborn determination never faltered in the face of all this adversity. Each disaster was a lesson, quite a costly one, from which he profited. In the mean time, the coarse fish population was diminishing sharply, a problem which had certainly played an important part in the decline of the fishery. Even during this lean spell, there were glimpses of the success which might be round the corner. In 1978 the chairman commented at the AGM, 'Marginally fewer trout had been caught than during the previous season, but the quality was significantly improved. Over 400 trout topped the two pound mark.' Then, in 1980, Lyng plumbed the depths, but that was the storm before the calm; the second phase had ended.

So to the third phase. After 1981 figures speak for themselves, with a return of about 80 per cent on fish stocked, even though this is quite a big lake with large areas of water way out of fishing range. Two pound trout are not even counted now, but I note that, in 1987, the Chairman counted more than 700 trout over three pounds and another 150 topped four pounds. I think that the biggest brownie was 8¼lb, the best rainbow somewhat larger. The real triumph is the unpredictability of the fishing; the angler never knows what is coming next, a modest three-quarter pounder or a monster ten times that size. Most of the fish are in beautiful condition, no tail-less wonders here, and the brown trout, especially those between two and three pounds, are super fish.

The Proof of the Pudding

Though much of the credit must go to Jack Fitt, running a fishery is a team game and we have a first division side. Netting and stocking the trout run like clockwork nowadays though, at one time, it used to be a shambles. The tradesmen in the club have done a wonderful job with the electrics, the building and the plumbing, others have helped enormously with the supply of materials. I dare not name them, for fear of leaving one out and giving offence. I cannot believe that there is a club in the country which offers its members such superb fishing and facilities at such modest cost; our aim is to be a genuine club, open to people in all walks of life who are willing to work together. The best example of the team spirit is probably the stocking programme. In 1986 there were twenty-three separate stockings, a job which needs a minimum of seven or eight helpers, but there were often more.

We enjoy terrific sport, but I would not have you believe that everything is perfect. We have our fair share of problems to contend with. For the last three years, we have had to suffer an absolutely foul-looking algal bloom. It appears some time in February when the water is at its coldest, and leaves the water brown and horrible until well after the start of the season. Then a couple of days of warm weather puts all to right almost overnight; as if by magic the water is crystal clear again. It's not toxic, which is just as well, though one or two sickly fish may succumb, but the trout certainly don't like it and go right off the feed for several weeks, which slows down our fish-rearing programme. We also have to live with the ever-present threat of eye fluke and proliferative kidney disease. The latter has killed a lot of our small fish during the last couple of years. These are losses which one just has to accept, but it is important to budget for them, so that replacement fish can be bought. Floods are another ever-present menace. After five or six quiet years the late summer and autumn of 1987 were very difficult. Our security is much better than it used to be, but we still lost quite a lot of small fish from one pen. If there is a gap anywhere, they can be guaranteed to find it.

Nor is the fishing itself without its problems. In common with many lake fisheries, sport is good until late June, when the water starts to get warm and the fly hatches diminish. From then on, the rise is much less predictable, though those who fish early morning or late evening still find a few fish rising. Generally, however, sport is slow and only those who fish with a well-sunk fly catch a few trout in the daytime. A large pipe was installed a few years ago, to bring in extra water from the river, mainly to help with our trout rearing. This seems to have helped the fishing as well, as sport seems a little more lively and not just in the close proximity of the pipe. Late stockings also help to keep the sport ticking over, but I doubt that this imbalance in the fishing can be completely overcome.

The prospects for the future are always uncertain; there is a lot of land available on the site and the owner has planning permission to develop a small complex of chalets and a club house. It's no good grumbling about development; without it, the lakes wouldn't have been there in the first place so we wouldn't have had any fishing at all. Without doubt, this development could have a big impact, as some of the holiday-makers would be coming for the trout fishing and a certain number of rods would have to be made available. Others would come for the excellent coarse fishing to be found in the smaller lakes, in fact the plan is for the chalets to be built near one of these. The biggest problem, as I see it, is

Fig 105 Evening fishing can be superb.

not to accommodate the extra fishermen, but to retain our good club spirit. It is the ability to work together which has made Lyng the superb fishery it is and the envy of all our visitors. If that spirit is lost, Lyng will become just another trout fishery.

10
A Final Word

What we are working for, in the final analysis, is lovely surroundings and clear water. Figs 106 to 110 illustrate the pleasures of success. To have seen a mayfly hatch occur is a bonus and so is the trout I caught, a wild fish which had never seen a pellet in its life. I only put my rod up to humour Jim Tyree while he was taking some of the photographs for this book. I saw the mayfly, which was resting on the willow herb, one thing led to another and I was ordered to catch a fish. I am normally an angler of monumental incompetence; I once rose thirteen fish in succession, never even touching the first dozen, and breaking off in the last as I unleashed an exasperated strike, followed by a few words of regret. On

Fig 106   *How can a trout be so stupid as to mistake the two?*

Fig 107   *A fish takes the fly.*

Fig 108   We struggle for a
while ...

Fig 109   ... before I reel it
in.

Fig 110   The result – a lovely wild fish.

169

this occasion, everything went right. I hooked and landed the first fish that rose, and Jim had his photographs with the minimum of fuss.

Pleasure is to be found in doing the work to improve the fishery and its environment. This is where people play the most important part. I hope I have mentioned words like friends, helpers, team and club spirit sufficiently often to make it apparent that it is not just the fishing that benefits from this programme of self help. If people are interested and involved, we all benefit from the team work and sense of achievement. My friend, Jim Knights, is an example of what I mean. He isn't even a member of the clubs which fish the river, but he is still pleased to come along and lend a hand. He makes out that I run some sort of chain gang but, however that may be, if you are lucky enough to have a few Jims among your friends, the work almost does itself. Your landowner will probably think that you are all raving lunatics to begin with but, as the improvements become apparent, even he risks infection.

Just a word of warning, however; this fishery improvement work is all well and good but, if you become too involved, you will suddenly realise that you are doing much less fishing than you used to do. It is fatal to go out taking both tools and fishing rod. It is a near-certainty that you will become engrossed in a job, and 'I'll just do a bit more' has suddenly taken over from the fisherman's more usual 'Just one more cast'. This has happened in my case, but I'm still not sure whether I mind or not. It may pay you to keep one day sacred to fishing, if you don't want to end up like me.

Finally, a more serious point. No amount of improvement work will bring back the fish, if the water is polluted. Our environment has taken a battering in the name of 'progress'. The more sophisticated we become, the greater seems to be our capacity to pollute. Fortunately, the conservation lobby is gradually gaining in strength and soon even politicians may become aware that there are votes to be won in this field. Most pollution can be overcome; the acid rain which we so generously export free of charge to other countries can be dramatically reduced, provided that governments are prepared to spend the necessary money. Water authorities have the ability to make sewage discharge perfectly harmless; again it's a question of capital expenditure. Farm pollution can most certainly be brought under control, though some of the damage already caused may be very long lasting. It is obvious, therefore, that we all have a responsibility to work for a cleaner future.

The first step is to support the bodies which already work on our behalf. I have frequently mentioned the Anglers' Co-operative Association.

Fig 111 'There's more to fishing ...'

I think that all clubs and individual anglers should be members. Anglers should all support the body that represents their particular branch of the sport which, for the game fishermen, is mainly the Salmon and Trout Association. We have a strong local branch, but the membership is pitifully small nationally. Once the individual associations are strong and active, then the National Anglers' Council can work more effectively.

Next, we should learn from such successful organisations as the British Association for Shooting and Conservation and the Royal Society for the Protection of Birds. Angling is light years behind these bodies in the promotion of our interest. The skill of the RSPB in the fields of public relations and publicity is absolutely outstanding; birds are a daily diet on television. We should try to take a leaf out of their book and promote angling in a positive way. Recently John Wilson made an excellent series of angling programmes for ITV and this is a big step in the right direction. It is high time that the best aspects of fishing were shown.

I believe that anglers should take a more active interest in the well-being of their sport and not expect somebody else to do all the work for them. I hope that I may have encouraged a few more of you to have a try and to adopt some of the suggestions made in this book. Once you start you will always be finding something extra to do. I hope you will find as much pleasure in it as I do!

Further Reading

Behrendt, A., *The Management of Angling Waters* (André Deutsch 1977)

Keith, A.C., *Trout in Norfolk* (Philip Allan 1936)

Plunket Green, Harry *Where the Bright Waters Meet* (André Deutsch 1983. First published Christophers 1924)

Sawyer, F., *Keeper of the Stream* (Allen & Unwin 1985. First published A & C Black 1952)

— *Nymphs and the Trout* (A & C Black 1981. First published Stanley Paul & Co 1958)

Many of the booklets published by the Institute of Fisheries Management contain useful and relevant information (see *Useful Addresses*).

Some of the above may be out of print, but all should be available through your library.

Pamphlets and articles

Bulletin Français de Pisciculture, Nos 258, 285

Le Moniteur, *Environment* (26 July 1982)

Montonati, J., *Travaux Aquatiques du Sud-Ouest*

Articles by Professor Faugère of the Laboratoire Municipal de Bordeaux

The above should be obtainable through a large reference library.

Useful Addresses

Anglers' Co-operative Association
23 Castlegate
Grantham
Lincolnshire
NG31 6SW

Institute of Fisheries Management
22 Rushworth Avenue
West Bridgeford
Nottingham
NG2 7LF

Nautex suppliers:
Thomas Mawer Ltd
Fleet Chambers
58 Jameson Street
Hull
HU1 3LS

Salmon and Trout Association
Fishmongers' Hall
London Bridge
London
EC4R 9EL

Siltex suppliers:
Needham Chalks Ltd
Needham Market
Ipswich
IP1 8EL

The Sports Council
16 Upper Woburn Place
London
WC1H 0QP

Index

175